THE NATURE OF
HIDDEN
WORLDS

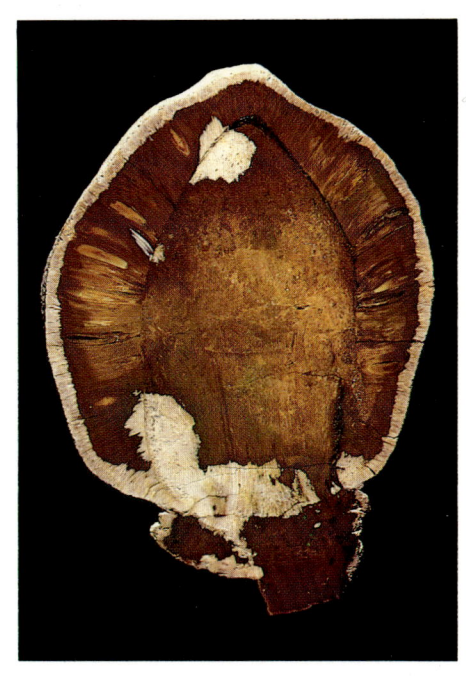

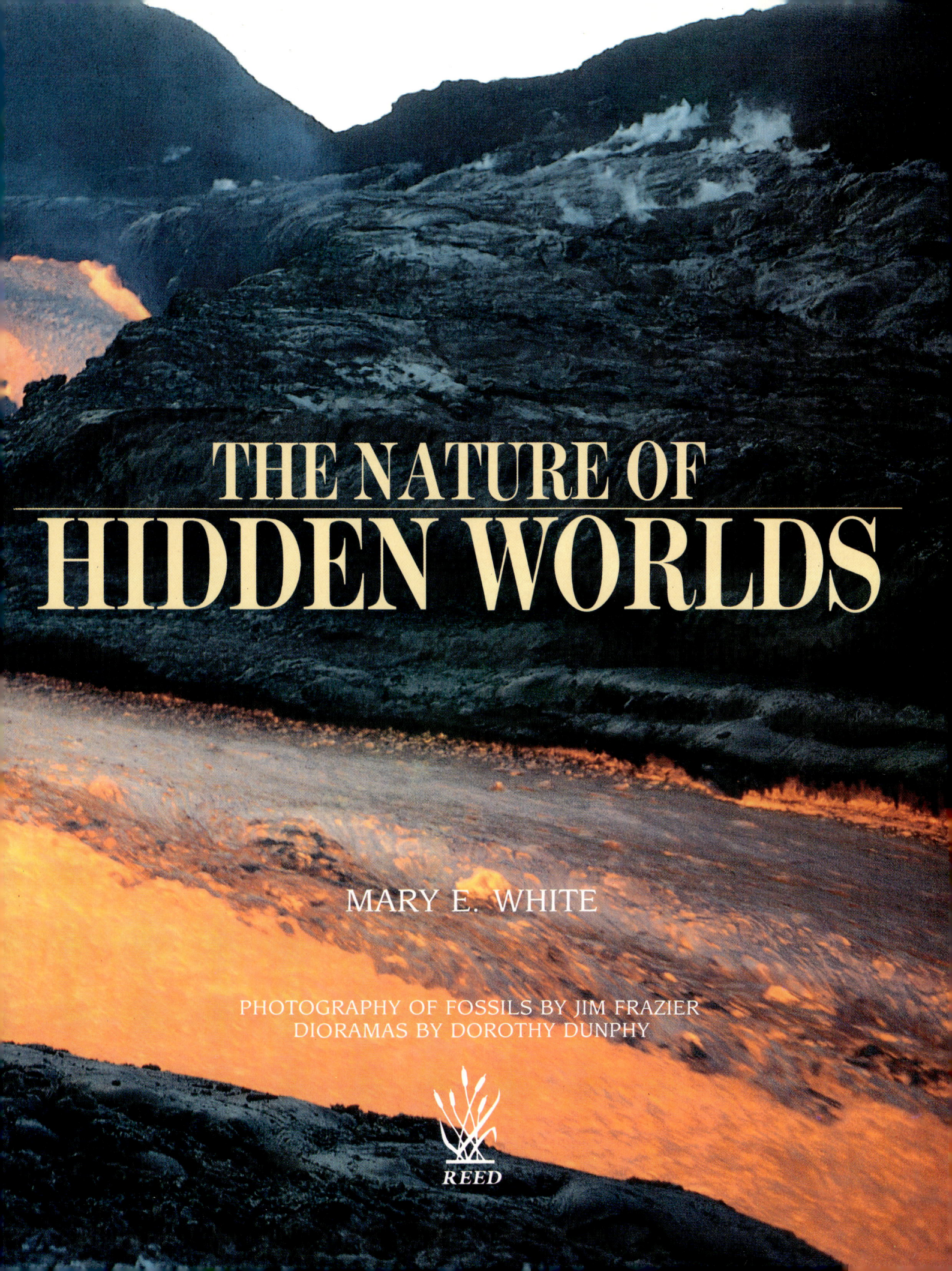

THE NATURE OF HIDDEN WORLDS

MARY E. WHITE

PHOTOGRAPHY OF FOSSILS BY JIM FRAZIER
DIORAMAS BY DOROTHY DUNPHY

REED

PREVIOUS PAGE:

A volcano erupts on Iceland and incandescent lava flows in a glowing river between black basalt banks. The Mid-Atlantic Ridge runs across Iceland, and the land is being torn apart. The volcanic activity is part of the dynamic cycle in which new crust is formed in our planet, when lands move and oceans grow.

OPPOSITE PAGE:

The Green Python, Chondropython viridis, is a nocturnal Rainforest species that lives in hollow trees or in the epiphytic Fern clumps growing on them. This specimen was photographed by Jim Frazier in the Iron Range in north-eastern Australia. The young of the species are lemon yellow, orange or gold in colour.

First published in 1990 by
Reed Books Pty Ltd
3/470 Sydney Road Balgowlah NSW 2093

National Library of Australia
Cataloguing-in-Publication Data

White, M. E. (Mary E.)
The Nature of Hidden Worlds

Includes index
ISBN 0 7301 0275 0

1. Paleobotany — Australia
2. Paleobotany — New Zealand
I. Title
561'. 1'0994

Produced in Australia by the Publisher
Edited by Helen Grasswill
Designed by Bruno Jean Grasswill
Proof-reading by the Publisher
Fossil photography by Jim Frazier,
   Mantis Wildlife Films Pty Ltd
Cartography by Bruno Jean Grasswill
Dioramas by Dorothy Dunphy
Additional illustrations by
   Bruno Jean Grasswill
Typeset in Australia by
Excel Imaging Pty Ltd
Printed in Singapore by
Times Offset Pte Ltd

ACKNOWLEDGEMENTS

I acknowledge with gratitude the contributions made by friends and colleagues to *The Nature Of Hidden Worlds*.

Jim Frazier and Dorothy Dunphy, whose fossil photography and art work respectively were so important in *The Greening Of Gondwana*, have collaborated with me again, and I would not have been able to achieve my goals without them. Once again, the Reed Books team of Bill Templeman, publisher, Bruno Grasswill, book designer and artist, and Helen Grasswill, editor and mentor, have contributed their special talents. I thank them for their dedication, interest and co-operation.

It would not have been possible for me to write *The Nature Of Hidden Worlds* had I not had the generous help and advice of a large number of earth scientists of different disciplines. Their encouragement has meant a great deal to me. In particular, Dick Evans of the University Of New South Wales has given generously of his time, and his enthusiasm has helped me through the rough patches. He kindly read the early manuscript and his criticisms were most helpful.

Source Of Fossil Specimens: The Australian Museum and the New South Wales Department Of Mines provided access to the majority of the fossils pictured in this book. My thanks for this privilege go to the Australian Museum Trust for permission, and to Alex Ritchie and Bob Jones for help and advice; and to John Pickett and Gary Dargan for access to the Mining Museum collections. John Bennett, a private collector from Perth, lent valuable *Pentoxylon* specimens, and the Western Australian Museum made specimens available for photographing in Perth. The geology departments of the University Of New South Wales and Macquarie University lent some special fossils, and John Pickett supplied thin sections of Corals.

Source Of Maps: Chris Scotese of Shell Research in Houston, Texas, USA, most generously gave permission for his computer programme (held by the Bureau Of Mineral Resources in Canberra) to be used in making polar projection maps showing the changing arrangements of land and sea through time. Steve Holliday at the Bureau made the printouts, and I thank the Bureau Of Mineral Resources for help in this and other matters. The palaeogeographic maps used in this book are based on the Earth Science Series of the Bureau, with their permission. The palaeogeographic maps of New Zealand are based on a series by Graeme Stevens of the Geological Survey of New Zealand, with his kind permission. Alberto Albani and Peter Rickwood of the University Of New South Wales supplied information on the seismic studies on bedrock topography which they and their co-workers have been carrying out on the continental shelf near Sydney.

Photography: In addition to Jim Frazier's fossil photography, I am most grateful for permission to use general colour transparencies contributed by the following people to illustrate this book: Peter Rickwood, Iceland volcanoes and rift; Densey Clyne, living plants, animals and landscapes; Bob Tingey, Antarctic landscapes; Roger Hocking, Western Australian geology and trace fossils; Ian Tyler, iron ore in Western Australia; Murray Johnston, Canning Basin reefs; Bureau Of Mineral Resources reference collection, geology; Alan Wells, sand dune deserts of the Centre; John Forlonge, South American vegetation; John Barrie, Pleistocene Vertebrate fossils; Tasmanian Education Department, pictures of Tasmanian scenery by Max Banks; Michael Greenwood, New Zealand landscape and divaricating plants; George Poinar, New Zealand fossils and vegetation; Hal Cogger, the Tuatara; and John Fields, Alex Ritchie, John Casey, Margaret Bradshaw, Ben Wallace, Pat Rich, P. Stewart, F. Coffa and S. Morton for various pictures. My thanks, too, for expert, specialised photography of micro-organisms contributed by: Ron Oldfield of Macquarie University, Conodonts, Diatoms, Radiolarians; Robin Helby, Consulting Palynologist, spores, pollen, Dinoflagellates; Sami Shaffik, BMR, Coccoliths; Denis Burger, BMR, early Angiosperm pollen; John Cleasby and Gunther Bischoff, Macquarie University, Foraminifera, Radiolaria, and Ostracods; Peter Jones, Ostracods; Helene Martin, Tertiary pollen; and Andrew McMinn, Dinoflagellates.

Many other people have also helped me by answering questions and supplying advice and encouragement. To everyone, my sincere thanks for your interest, help and support.

Mary E. White

PREFACE

What did the dead land-world look like when the first Amphibians peered at it from the edges of their watery habitats? Through what green and lovely wilderness did Dinosaurs roam, browsing on the tree-tops? When the first Ape-men walked upright on the savannah, was their world very different from ours?

Finding the answers to these and a multitude of other questions about the distant past is an interesting and rewarding exercise for the mind. But it is far more than that. To know and understand a little of Earth's history imparts a new perspective. We see the three million years of evolution of the human race as insignificant against the background of the enormous span of geological time, and note the irony of the situation in which we — who see ourselves as the pinnacle of the evolutionary process — are blindly bent on destroying the biosphere. Knowledge of the distant past also gives us a new perspective on problems facing the world today: it reminds us of the natural laws which have governed the living planet since the beginning, shows us the directions we must follow if we are to survive, and warns us that extinction is ultimately the price paid for rapid environmental change.

How the secrets of past ages can be unravelled is itself an interesting story, for the facts about the changing environments through geological time and about the evolution of plants and animals are hidden in the rocks of the Earth's crust. It requires the cumulative evidence from many disciplines of Earth Science to obtain comprehensive knowledge of the physical environment in prehistoric times and information from the Fossil Record to visualise the flora and the fauna.

An ever-changing picture emerges of what it must have been like to be alive when prehistoric animals and plants in their turn were the inhabitants of the Earth. We see evolution at work; its driving force environmental change, its agent Natural Selection.

You and I, and all living creatures, have our origins in the primaeval oceans. Our embryonic children show remnant links with our fishy ancestors in the gill slits which scar their necks in the early stages of their development while they are swimming in their own small, private seas. So the journey back through time and down the evolutionary ladder is in fact our own journey, and allows us to find our beginnings and come to know what our place is in the overall scheme of things. We can see our human situation in context with the total environment of the living world, and realise that the laws of Nature which governed the life, death and final extinction of plants and animals throughout geological time are still the laws governing us and our modern world. These laws are essential in maintaining a balanced biosphere — and if balance is lost, the planet dies.

Diatoms are important microscopic components of the marine phytoplankton whose photosynthetic activities contribute to the balancing of gases in the Earth's atmosphere by taking up carbon dioxide and releasing oxygen. They are unicellular Algae (members of the Plant Kingdom) with shells of silica, and they occur in astronomical numbers in cold seas — and they have done so through the last 70 million years of geological time. Their delicate perfection epitomises the intricacies of the connection between life and the inanimate in the biosphere. As living entities they help to make the air fit to breathe, while the mineral residue of their shells contributes silicon to the sediments that accumulate on the sea floor to form rock over time. In the modern world the sea floor in Antarctic and Arctic regions is blanketed with shells of Diatoms.

RON OLDFIELD

The Koala, one of Australia's most publicised and best-loved animals, is fast becoming an endangered species in the wild. It has been fighting a losing battle against the human species for the last 50,000 years. First the Australian Aborigines found it easy prey, killing it in the trees; then the Dingo introduced by these early people hunted it on the ground. Fire used in hunting took its toll, and the limited diet, which is restricted to a few species of Eucalypt, restricted it to specific areas. Then came white settlers, killing for skins, for meat, for fun, and progressively destroying its habitats. Protection came eventually, but arguably too late. The stress of modern living for these ancient and highly specialised creatures now makes many populations susceptible to infections that affect fertility and cause blindness.

The human species has been able to reach plague proportions, to alter the global environment, to upset the balances . . . and, so far, to survive. By sheer weight of numbers we have changed the face of the Earth, believing we have a God-given right to do so and disregarding the rights of other living things with which we share the planet. To date our cultural and technological evolution has shielded us from the dire effects of flouting basic natural rules. Nevertheless, we have put the very survival of Earth as a living entity at risk. We have no way of knowing when the scales will tip and the slide towards extinction will become irreversible. It is time to stop and read what is written in the rocks and heed the warnings implicit in the evidence from the remote past.

Mary E. White
Sydney, Australia

KEY TO PALAEOGEOGRAPHIC MAPS

MARINE ENVIRONMENTS

Shallow sea

evaporites

carbonates

reefs

glacials

Deep sea and oceanic

CONTINENTAL ENVIRONMENTS

sediments spread by rivers or wind

land undergoing erosion

glacial

peat swamps

mountains

VULCANICITY

eruption of flood basalts

central-type volcanoes

GENERAL INFORMATION

The **time scale** used in this book follows the 1983 Geological Time Scale in the *Decade of North American Geology* produced by the Geological Society of America and published in *Geology*, September 1983, page 504.

The series of **palaeogeographic maps** are based on the "Earth Science Series" of the Bureau of Mineral Resources, Canberra, compiled by G. Wilford, and adapted with permission. The north-south orientation of the continent shown in this book follows the palaeomagnetic information for Australia used in compiling these earth science maps.

The **global reconstructions** showing North and South Polar projections are those of Dr Chris Scotese and have been produced by the Bureau of Mineral Resources from his computer software, with his permission.

In descriptions of palaeogeography, the points of the compass are used as they refer to the modern continent, i.e. if the seas are said to "retreat to the east", that is the east as it is now.

The captions for the **fossil specimens** include abbreviated notations of where they are held, as follows:

AMF	The Australian Museum, Sydney
MMF	The Mines Department Collection, Sydney
U.NSW	University of New South Wales (Geology Department), Sydney
MQU	Macquarie University (Geology Department), Sydney
BMR	Bureau of Mineral Resources, Canberra
WAM	Western Australian Museum, Perth
MV	Museum of Victoria, Melbourne
JD	Jack Douglas, Melbourne

A **glossary** on pages 242–247 explains terms and words which may not be familiar to all readers.

CONTENTS

A DUNE DESERT IN THE AMADEUS BASIN
Deserts with dunes of wind-blown sand have
been a feature of the dry interior of the
Australian continent for the last 300,000
years. This amazing landscape of wind-
sculptured red sand lies between Finke and
Crown Point in the Northern Territory.

The Age Of Ancient Life
(The Palaeozoic Era)

ALAN WELLS

INTRODUCTION

Hidden in the rocks of the Earth's crust is a record of the evolution of our planet. It tells the story of an ever-changing world through the enormous expanse of geological time and of the life which transformed it into a living planet.

This story is there for us to read. Studying the structure of the rocks enables us to reconstruct the environments of the hidden worlds of the past, and examining the fossils they contain reveals the "Nature" — the plants and animals — which lived in the changing environments of successive geological Periods.

The history of the Earth goes back to its birth about 4600 million years ago. The earliest times, before there was life in its waters, are obscure. It was a dead planet without free oxygen but it contained elements which would eventually combine to form life. It was the activities of living organisms with the power to photosynthesise — the distinctive characteristic of plants — which gradually created the living, oxygenated world in which we and all the living things around us have evolved. Early life was microscopic and remained primitive until the environment had evolved to a stage where it presented new opportunities for experimentation.

Scientists divide geological time into two Eons, the **Cryptozoic** and the **Phanerozoic**. These Eons are differentiated by the absence or presence of living things whose requirements for life were similar to those of creatures alive today and which left a visible Fossil Record.

The **Cryptozoic Eon**, or Age Of Invisible Life, encompasses the first 4000 million years. (As modern research advances, more and more life is being detected in this slice of Earth history.) During this Eon the global environment evolved in preparation for the life-forms which now are characteristic of the planet.

The **Phanerozoic Eon**, or Age Of Visible Life, covers the last 570 million years of geological time, and therefore encompasses only a small fraction of the Geological Time Column. It is the environments of this Eon which form the main subject of this book, insofar as they can be interpreted for Australia, and for New Zealand from the time when it emerged from under the sea some 150 million years ago.

Throughout *The Nature Of Hidden Worlds* the aim is to establish the conditions, circumstances and influences surrounding the plants and animals which lived in the geological Periods and to see their evolution against this environmental background.

Spriggina floundersi, one of the world's oldest fossils of a multicellular animal, lived about 630 million years ago, before the Cambrian Period which starts the Age Of Visible Life. Spriggina is a member of the Ediacara Fauna from the Flinders Ranges in South Australia and is highly organised for such an early animal, having a segmented body and a distinct head. It was a soft-bodied animal whose relationships are uncertain. Ediacaran animals were possibly "experimental", constructed on a different plan to the multicellular animals (Metazoa) that appeared later. (Specimen AMF. Magn.X 4.7)

GEOLOGICAL PERIODS OF THE PHANEROZOIC EON

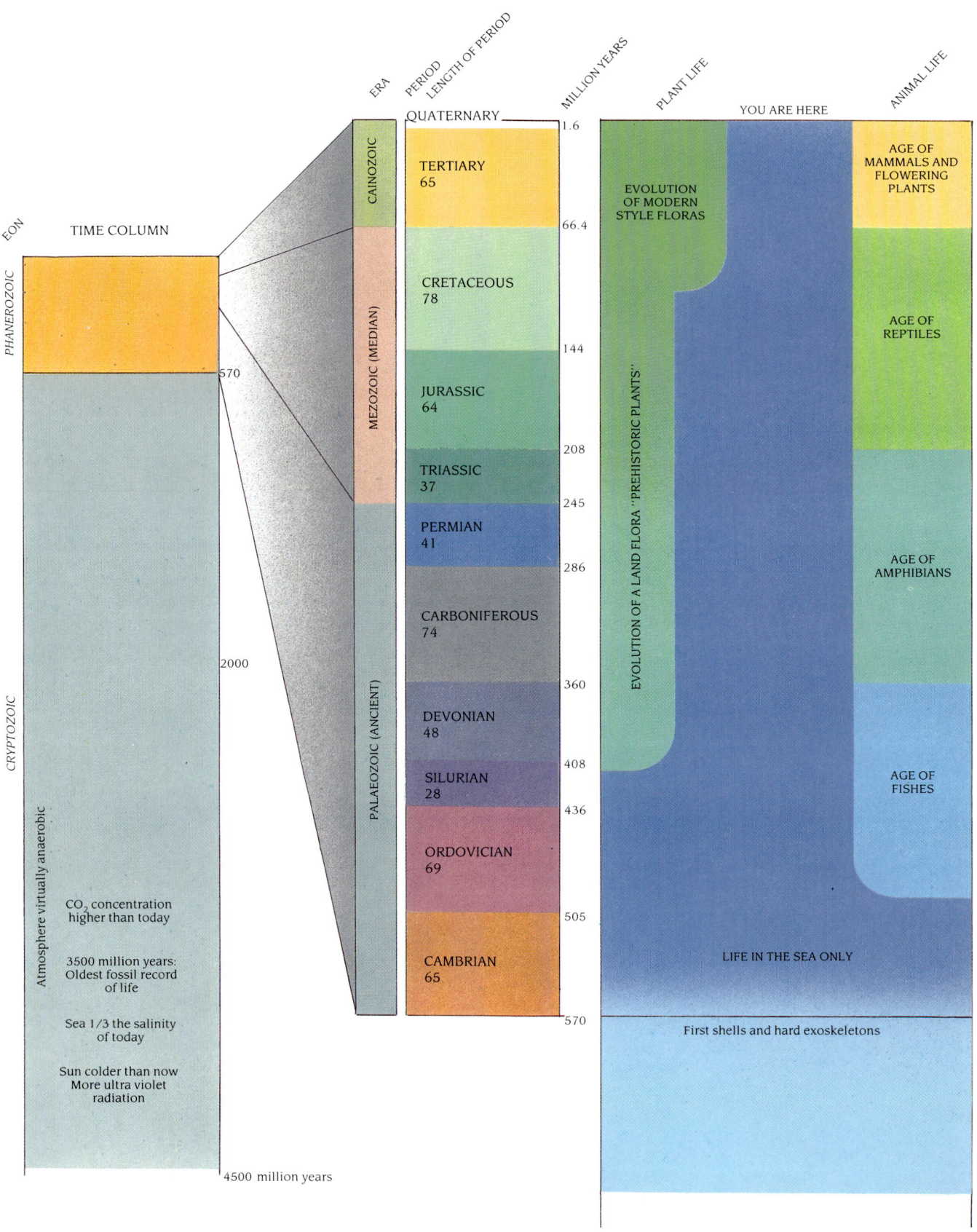

CLASSIFICATION OF LIVING ORGANISMS — THE FIVE KINGDOMS

The Animal and Plant Kingdoms are major units of classification, but there are three other kingdoms — those of Fungi, Protista and Bacteria.

Plants have the green pigment chlorophyll which enables them to photosynthesise food materials and thus they are the basis of food chains for most other living things. Apart from the single-celled, and some multi-celled, Algae, plants are immobile.

Animals are usually mobile, and their lack of chlorophyll renders them ultimately dependent on plants for food.

Fungi are regarded as a separate Kingdom because they lack the ability to synthesise food, being either saprophytic (living on dead organic matter) or parasitic (living on other living organisms). They are mostly immobile, like plants.

It is now generally accepted that the multitudinously abundant, microscopic "single-celled" organisms which used to be assigned to the Animal Sub-kingdom Protozoa are best regarded as a separate Kingdom, the **Protista**. These organisms are not merely single "cells" of which other organisms are made. Each unit is a complete individual which carries out the many functions of multi-celled organisms. Protista are now described as "acellular" rather than unicellular.

Many Protists are intermediate between plants and animals in their characteristics. The only Protists which have a Fossil Record are those which produced shells or outer cases, or skeletal structures. The microscopic blobs of living protoplasm, of which many of them were composed, left no record. Prominent examples of Protists with a Fossil Record are Foraminifera and Radiolaria, the former with shells and the latter with intricate skeletons.

Bacteria are of ultra-microscopic size and are regarded as a separate Kingdom because they have characteristics different from those in any of the other Kingdoms.

THE LINNAEAN SYSTEM OF CLASSIFICATION OF PLANTS AND ANIMALS

Plants and animals are classified according to the system instituted by the eminent Swedish scientist, Linnaeus, in 1753. In his "binomial system" each recognisably different plant and animal is given a generic and a specific name. Closely related individuals belong to the same *genus,* and minor differences create *species* within the genera. Genera are grouped into **families**, and families into **orders**. Orders, in turn, are grouped into **classes**, and classes into **phyla** (in the Animal Kingdom) or **divisions** (in the Plant Kingdom).

The Tasmanian "Soft Tree-fern" (Dicksonia antarctica) is classified thus:

Plant Kingdom
Division: Pteridophyta
Class: Filicales
Order: **Polypodiales**
Family: DICKSONIACEAE
Genus: Dicksonia
Species: antarctica

The frilled-necked Lizard (Chlamydosaurus kingii) is classified thus:

Animal Kingdom
Phylum: Chordata, section Vertebrata
Class: Reptilia
Order: **Squamata**, *suborder* **Sauria**
Family: AGAMIDAE
Genus: Chlamydosaurus
Species: kingii

CLASSIFICATION OF THE PLANT KINGDOM

The Plant Kingdom is divided into four major phyla:

1. The Thallophytes

ALGAE

Organisms which are classified as plants because they have chlorophyll and the ability to photosynthesise. There are seven classes of Algae. Seaweeds and some freshwater forms have a simple plant body composed of matted filaments, lacking true roots, stems and leaves and without conducting tissue. They reproduce by spores and have an alternation of generations, requiring external water for their free-swimming gametes. Their reproductive structures are usually single-celled. The green scum of freshwater pondweed consists of filaments of single rows of cells. There are single-celled Algae which are mobile, as are some small colonial forms. Diatoms, the abundant organisms in plankton, secrete an external case made of silica, and other microscopic Algae make calcareous shells

Class Chlorophycophyta
Green Algae, green seaweeds
Clas Phaeophycophyta
Brown Algae, brown seaweeds
Class Rhodophycophyta
Red Algae, red seaweeds
Class Charophycophyta
Stoneworts which secrete lime
Class Chrysophycophyta
Diatoms
Class Pyrrophycophyta
golden brown, Dinoflagellates
Class Euglenophycophyta
euglenoids

2. The Bryophytes

Non-vascular plants which reproduce by spores and have an alternation of generations

Class Hepaticae
the Liverworts
Class Musci
the Mosses

A Liverwort, *Marchantia berteroana*. Liverworts and Hornworts are Bryophytes which differ from Mosses in the nature of their reproductive structures and in their growth habits.

A Moss, *Funaria hygrometrica*. Mosses are Bryophytes, plants without vascular tissue. They reproduce by spores borne in capsules. In this species the capsules are coloured. Bryophytes have a long fossil history which probably dates back to the first land-plants about 400 million years ago.

3. The Pteridophytes

Vascular plants which reproduce by spores and have an alternation of generations. Free water is required for the transfer of gametes to effect fertilisation

Class Filicales
Ferns (Bear spores in sporangia on the under surface of their leaves. An ancient plant group going back about 300 million years.)
Class Psilophyta
Psilophytes, the first land-plants (Some 420 million years ago they made the transition out of the water, enabling animals to move out onto the land as well. There were two lines of evolution, one via Zosterophylls to Clubmosses, and the other via Rhyniophytes and Trimerophytes to Ferns, Horsetails and all seed-bearing plants. Only two genera and four species of Psilophytes are alive today.)
Class Sphenophyta
Horsetails (Plants with segmented stems, leaves in whorls at nodes, terminal cones bearing spores, and an alternation of generations. They were among the early land-plants and grew as trees in Palaeozoic swamp forests. Thereafter, they grew like rushes and reeds on water margins. Today only one genus, *Equisetum*, still exists of this ancient plant group.)
Class Lycophyta
Clubmosses, also known as "Fern-allies" and tassel-ferns (*Baragwanathia*, a fossil found in Australia, is among the oldest land-plants, 420 million years. During Early Palaeozoic times Clubmosses were a major component of swamp forests, growing as very tall trees. All living members of this group are herbs which require damp habitats.)

Tree-ferns, *Cyathea australis* and *Dicksonia antarctica*, whose ancestry goes back through about 300 million years.

A fertile Fern pinnule, showing the aggregations of sporangia (sori) on the lower surface. The spores which are produced in the sporangia germinate to produce a prothallus, a small leaf-like plant which forms the alternate generation. The prothallus produces male and female cells (gametes) which fuse to form a zygote from which the Fern plant grows.

A Horsetail, *Equisetum sp.* Horsetails have segmented stems with leaves which form sheaths at the nodes (in this species the leaves are reduced to small teeth). They reproduce by spores borne in terminal cones and there is an alternation of generations in the life cycle. Only one genus, *Equisetum*, survives today from this group of plants which was so important and abundant in the Fossil Record. Living representatives are all small or medium sized herbs, but in the Palaeozoic swamp forests Giant Horsetails grew as large trees with the Giant Clubmosses.

A Psilophyte, *Psilotum nudum*, whose leafless stems have a strand of vascular tissue. The yellow sporangia are borne in clusters of three in the axils of small scales. Psilophytes were among the first plants to colonise the land 420 million years ago and were ancestral to the Ferns, Horsetails, and all seed-bearing plants. Today only two living genera of Psilophytes survive. *Psilotum* is distributed worldwide, and *Tmesipteris*, which grows as an epiphyte on Tree-fern trunks, is restricted to Australia and Polynesia.

A Clubmoss, *Lycopodium sp.* Modern representatives of this ancient plant group which was established 420 million years ago are herbaceous ''Fern allies'', limited to specialised damp habitats. In the geological past there were Giant Clubmosses which grew as large trees in swamp forests.

19

4. The Spermatophytes

Seed Plants — vascular plants which reproduce by seeds (The development of seeds eliminated the alternation of generations and severed the ties with the water habitat, which were the legacy of land-plants' aquatic algal ancestry.)

Class Gymnospermae

Seed Plants in which the seeds are not enclosed in a vessel

Order Pteridosperms

Seed-ferns (extinct)

Order Cordaitales

the Cordaites (extinct), ancestral Conifers

Order Glossopteridales

the Glossopterids, Permian plants of Gondwana, probably ancestal to several groups of higher plants

Order Cycadophyta

the Cycadeoids (extinct) and Cycads (living)

Order Ginkgophyta

the Ginkgos or Maidenhair Trees (an ancient plant group whose members were important constituents of forests from about 260 milllion years ago. Their importance declined during the Mesozoic Era, and today only one species survives.)

Order Coniferophyta

the Conifers (An ancient plant group which dominated world floras in the Mesozoic. While Flowering Plants took over the dominant role in most vegetation types from the Cainozoic onwards, Conifers are still dominant in some climatic zones, remaining important components of vegetation. Seeds are borne in cones.)

Cycads, Palm-like plants which are members of an ancient group of gymnospermous Seed Plants with a long fossil history starting about 260 million years ago.

A Ginkgophyte, Ginkgo biloba. The "Maidenhair Tree" is the last survivor of a plant group which dates back to Permian times about 260 million years ago.

Class Angiospermae

Flowering Plants or Angiosperms (In the Angiosperms, the seeds are enclosed in a vessel, there is double-fertilisation resulting in the production of endosperm — extra food for the embryo — in the seed, and the reproductive parts are arranged in flowers. Flowering Plants evolved only about 120 million years ago, and during the Early Tertiary they became dominant in most vegetation types.)

Order Monocotyledonae

the Monocotyledons (Flowering Plants in which there is one seed-leaf in the embryo, leaves usually have parallel veins, and the vascular tissue in the stems is in the form of many separate strands. Grasses, Lilies and Palms are examples.)

Order Dicotyledonae

the Dicotyledons (Flowering Plants in which there are two seed-leaves in the embryo. Leaves are net-veined. Secondary wood forms a cylinder, and the wood elements comprise vessels which are a type of wood cell absent from Conifer wood. Gum trees, Banksias, Oaks, Daisies and Roses are examples.)

CLASSIFICATION OF THE ANIMAL KINGDOM

The Animal Kingdom is divided into twelve major phyla:

1. Phylum Porifera

SPONGES
Primitive organisms with no nervous system or sense organs. Adults are attached and motionless. They filter water through the pores and canals in their structure, propelling water currents with special cells with cilia (beating hair-like projections). Sponges have a skeleton of needle-like spicules

ARCHAEOCYATHIDS
Ancestral Sponges confined to the Cambrian Period

Skeleton of a Glass Sponge. In their tissues Sponges have needle-like spicules composed of lime or silica. These spicules in the deep-sea Glass Sponges are made of opaline silica and they are rigidly united to form a lattice. The skeleton so produced is protected against torsion by the development of spirals of silica which girdle the structure at 45° to its axis. Glass Sponges can be traced back to the Cambrian Period, 550 million years ago, and there were many types of them flourishing in the Late Palaeozoic. The major extinction event at the end of the Permian resulted in most of the Glass Sponges disappearing, but the group had recovered by the Jurassic Period when there was evolution of many new forms. They were important members of the fauna during the Cretaceous, and some forms remained abundant on continental shelves during Tertiary times. Today Glass Sponges are comparatively rare, and those which survive are mainly deep-sea forms.

2. Phylum Coelenterata *also known as Radiata or Cnidaria*

Class Hydrozoa
Hydroids
Class Scyphozoa
Jellyfish
Class Anthozoa
Corals
Sub-class Alcyonaria
Soft Corals
Order Pennatulacea
Seapens
Sub-class Zoantharia
Sea Anemones

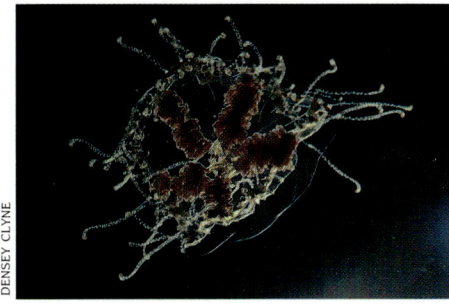

Jellyfish. Phylum **Coelenterata**, class Scyphozoa.

Coral. Phylum **Coelenterata**, class Anthozoa.

A Sea Anemone. Phylum **Coelenterata**, class Anthozoa, sub-class Zoantheria. An attendant Anemone Fish is seen inside the ring of tentacles where it lurks, immune to its host's stinging cells, waiting to feed on scraps from the Anemone's meals.

3. Phylum Platyhelminthes *and*
4. Phylum Nemertinia

WORMS
Virtually absent from the Fossil Record because they have no hard parts

5. Phylum Mollusca

Class Monoplacophora
Very small-shelled ancestral Molluscs of Cambrian age
Class Polyplacophora
Chitons
Class Gastropoda
Gastropods
Class Pelecypoda
Bivalve Molluscs
Class Scaphopoda
Tusk Shells
Class Cephalopoda
Squids, Octopi, Ammonites, Nautiloids

A Nudibranch (Sea Slug). Phylum **Mollusca**, class Gastropoda. The brightly coloured Nudibranch is attacking a Soft Coral (Phylum **Coelenterata**, class Anthozoa, sub-class Alcyonaria).

A Garden Snail and a Diamond Slug. Phylum **Mollusca**, class Gastropoda. Land Snails and Slugs are the only Molluscs to make a transition from water to land life.

21

6. Phylum Annelida
   SEGMENTED WORMS
      Marine Worms, Earthworms and
      Leeches; many marine forms produce
      calcareous tubes

7. Phylum Onychophora
   Peripatus is the sole survivor of this
   ancient phylum

8. Phylum Arthropoda
   ARTHROPODS
      Creatures with jointed legs
   Sub-phylum Trilobita
      the Trilobites
   Sub-phylum Chelicerata
      Spiders, Scorpions, Ticks, Mites
   Class Merostomata
      Horseshoe Crabs
      Sub-class Eurypterida
         the Eurypterids
   Class Arachnida
      Order Scorpionida
         Scorpions
      Order Araneae
         Spiders
      Order Opiliones
         Daddy-longlegs, Harvestmen
      Order Acarina
         Ticks, Mites
   Sub-phylum Mandibulata
   (i) Aquatic Mandibulates
   Class Crustacea
      Sub-class Ostracoda
         Ostracods
      Sub-class Cirrepedia
         Barnacles
      Sub-class Malacostraca
         Lobsters, Crabs, Shrimps
   (ii) Terrestrial Mandibulates
   Class Myriapoda
      Centipedes, Millipedes
   Class Insecta
      Insects
         Order Ephemerida
            Mayflies
         Order Odonata
            Dragonflies
         Order Dictyoptera
            Sub-order Blattoidea (Cockroaches)
            Sub-order Mantoidea (Phasmid,
            Mantis)
         Order Coleoptera
            Beetles
         Order Hemiptera
            Bugs, Cicadas, Leafhoppers
         Order Isoptera
            Termites
         Order Orthoptera
            Grasshoppers
         Order Hymenoptera
            Ants, Bees, Wasps
         Order Mecoptera
            Scorpion Flies
         Order Diptera
            Flies, Mosquitos
         Order Lepidoptera
            Moths, Butterflies

A Scorpion. Phylum **Arthropoda**, *sub-phylum Chelicerata, class Arachnida, order SCORPIONIDA.*

A Hermit Crab in a Mollusc shell. Phylum **Arthropoda**, *sub-phylum Mandibulata, class Crustacea, sub-class Malacostraca.*

A Millipede and a Centipede. Phylum **Arthropoda**, *sub-phylum Mandibulata, class Myriapoda.*

A Cockroach. Phylum **Arthropoda**, *sub-phylum Mandibulata, class Insecta, order DICTYOPTERA, sub-order Blattoidea.*

A Cicada. Phylum **Arthropoda**, *sub-phylum Mandibulata, class Insecta, order HEMIPTERA.*

A Scorpion Fly. Phylum **Arthropoda**, *sub-phylum Mandibulata, class Insecta, order MECOPTERA.*

9. Phylum Ectoprocta
BRYOZOANS Moss animals

---

10. Phylum Brachiopoda
LAMP SHELLS

---

11. Phylum Echinodermata
ECHINODERMS
Starfish, Sea-urchins, Sand-dollars
Sub-phylum Pelmatozoa
Crinoids
Sub-phylum Eleutherozoa
Class Asteroidea
Starfish
Class Holothuridae
Sea Cucumbers
Class Echinoidea
Urchins
Class Ophiuroidea
Brittle Stars

A Sea Urchin. Phylum **Echinodermata**,
sub-phylum Eleutherozoa, class Echinoidea.

A Starfish. Phylum **Echinodermata**,
sub-phylum Eleutherozoa, class Asteroidea.

*A Sand Dollar makes its trackway in the
sand as it moves about feeding. Phylum*
Echinodermata, *sub-phylum Eleutherozoa,
class Echinoidea.*

A Crinoid (Feather Star). Phylum
Echinodermata, *sub-phylum Pelmatazoa.*

12. Phylum Chordata
Animals with backbones
INVERTEBRATE PROTOCHORDATA
Ascidians
Pterobranchia
Graptolithina
Graptolites
VERTEBRATES
FISH
AMPHIBIANS
REPTILES
BIRDS
MAMMALS
Marsupials, Monotremes, Placentals

PART ONE

UNDERSTANDING ENVIRONMENT

THE COMPLEX ENTITY

JIM FRAZIER

PLAGUES OF MOUSEKIND AND MANKIND

When a plague of an animal species occurs in Nature, it is a short-lived phenomenon. Famine and disease restore the balances and reduce the numbers to a level where food supply and space can sustain them. But the human species knows no such natural constraint and swarms in plague numbers, upsetting the balance of the biosphere.

''Environment'' has become a catchword of the last decade. Too often, though, our perception of its meaning is confined to the most obvious aspects of our immediate surroundings. The full concept of environment is immensely complex because it is the sum of all the systems, animate and inanimate, which make up the living world, and because it is dynamic.

To define **environment:** it is *a complex entity comprising all the conditions, circumstances and influences surrounding and affecting the development of an organism or a group of organisms.* (The ''group of organisms'' can be those in a specified habitat or the complete biota of the Earth.)

The global environment, which results in Earth being a balanced biosphere, is the sum total of all the smaller environmental units which comprise the whole. The entire Earth can be regarded as the largest living organism in our solar system . . . our friendly green planet, lonely in the vast sterility of space. The concept of Earth acting like a single living organism, maintained in equilibrium by the interaction of all its parts and systems, is not new. It is the basis of the Gaia, or Earth Mother, principle which takes this sort of thinking to its ultimate philosophical length. But whether one interprets our world in theological, philosophical or scientifically analytical terms, the inescapable fact remains: when the activities of Technological Man are excluded, the Earth indeed seems to have been *a balanced biosphere, actively maintaining and controlling conditions so as to provide an optimum environment for life.*

Through the long ages of geological time, the environment has looked after itself and maintained its balances. Now, for the first time, the multiplication of a species (our human species) has proceeded unchecked and has come to threaten the health and well-being of the planet.

Until the arrival of man, a natural balance had been maintained between populations and their environments. When an animal species reached plague proportions because the balance was temporarily lost, the environment was unable to sustain the numbers. Famine and disease reduced the swarming hordes to the point where the survivors, if there were any, could be fed and sheltered. We are familiar with the recurring plagues of European mice in the wheatlands of Australia, where the balance has been upset by agriculture and the introduced species breeds unchecked when winters have been warmer than usual and the food supply has been enhanced. When the food runs out, starvation, cannibalism and disease reduce the numbers. Another example is a problem which occurred in parts of Africa where zebra plagues followed the removal of lions from various territories by farmers who had deemed their cattle to be at risk. The zebras created a far more serious problem by depleting the grazing lands. Such cases are representative of the multitudes which could be cited.

The consequences of upsetting natural balances by causing rapid environmental change are obvious . . . and man-made changes are excessively rapid when seen in perspective against the relatively slower changes that have occurred over geological time, which themselves were often too rapid to be accommodated and led to extinctions.

THE DRIVING FORCE OF EVOLUTION

Charles Darwin wrote in *On the Origin of Species*, published in 1859, that in the isolated and unique environment which is the Galapagos Islands: "Both in space and time, we seem to be brought somewhat near to that great fact — that mystery of mysteries — the first appearance of new beings on this Earth." He recognised in those isolated ecosystems, which were so startlingly different and like clear, simplified pictures, the driving force of evolution. The inspiration he found in those remote islands, where he saw that the animals had adapted to the unique local circumstances and had become new species in the process, led finally to the propounding of his Theory Of Evolution. His concepts were so far ahead of their time that they were considered blasphemous.

In rare instances, as in the Galapagos today, one can clearly see the basic truth that *it is environmental change, through Natural Selection, that drives evolution*. It has always been so, since the inception of life on this planet.

The mutual interdependence of living things and their environment is abundantly clear as we look around us. It is also obvious that the interaction between the biota and its inanimate surroundings results in mutual modifications. When a change occurs in an individual or a population which renders it better able to cope with the new conditions imposed by an environmental change, Natural Selection will promote the better adapted forms. However, the original modification may be no guarantee of permanent success. When conditions alter again, a new adaptation might be better suited and consequently selected. Adaptations

A juvenile Tuatara (Sphenodon punctatus), New Zealand's "Dragon", is a disappointingly tame-looking descendant of the Dinosaur or Giant Lizard ancestor that was a member of the Gondwanan fauna living in Ancestral New Zealand when it separated from the supercontinent 80 million years ago. This small Dragon, admirably adapted for its life in modern New Zealand, is the end-product of a long process of selection of ancestors with characters best fitted to the changed environmental circumstances of their times.

HAROLD COGGER

JIM FRAZIER

Competition for light in a Closed Forest environment results in the evolution of life styles which give plants access to better light high in the canopy. Thus epiphytes like Staghorn Ferns and Orchids grow on the trunks and branches of trees. The seedlings of a Strangler Fig employ a similar strategy. They germinate in the fork of a tree where there is enough light for vigorous growth. As the plant grows it sends long roots down to the ground, enveloping its host and eventually strangling it by enclosing it entirely within its tissues.

which result in over-specialisation are dead-end modifications; open-ended adaptations which help the organism to cope in the short-term without prejudicing its long-term prospects are more likely to be successful. Thus some highly specialised organisms with very specific requirements are adversely affected by change, while other less specialised forms can tolerate a wide range of conditions and can adapt to gradual changes. In situations where there has been no real alteration in conditions for very long periods the genetic ability to adapt may decline, so that when change does occur the lack of diversity means that there are no variants suitable for selection. Species and populations showing this lack of vitality are replaced by others with better genetic potential for change.

Just as there is an inbuilt obsolescence in every living thing — which has a time to live, hopefully to thrive and breed, and then to die — so, too, does the same rule apply on the larger scale. Extinction is as much a part of the scheme of things as is death. Efficiency under prevailing conditions determines which organism will dominate, competition keeps each individual at peak efficiency, innovation is essential for survival if conditions change, and those that cannot adapt will become extinct. All these basic rules apply as well today as they did to Dinosaur evolution and extinction in the distant past. It is the alarming acceleration of the processes because of the "unnatural" entry of humans into the equation which is now the reason for concern. There is no time for adaptation, and species simply become extinct. In this context "time" has to be of "geological" dimensions, not measured in months and years.

When we look at the geological record, the great lengths of time involved can make the decline and final extinction of a group of plants or animals appear to have been sudden, though it might have taken a million years or more. Even one million years is a long time in terms of generations of living organisms.

Unfortunately the Fossil Record, which provides us with our information about past floras and faunas, is full of gaps — in fact, it might be best to regard it as predominantly gap with only isolated intervals filled in by knowledge. Chance played a part in selecting which organisms appear as fossils, and erosion has removed much of the story which could have been told by the rocks.

By studying the fossils of successive geological Periods we can see evolution at work. Adding a background of the local and global environments of the times, and assessing the environmental pressures on the flora and fauna, enables us to see the reasons for some of the evolutionary jumps which characterise the Fossil Record. When the sparse collection of facts which can be elicited from all sources to form a picture of the environment of a particular time are assembled, they can be no more than a generalisation. However, it helps to illuminate a small spot in the general darkness of the past which surrounds it. A shifting mosaic of local environments, vegetation types and animal communities was the pattern in the past, just as it is today, and most of them disappeared without trace.

Consider for a moment the last 3 million years, which form an insignificant fraction of the Geological Time Column but one for which there is a lot of information. We see the vast changes wrought by the Pleistocene ice age. It changed climates and altered landscapes worldwide. It caused extinctions among the fauna, and the origin of humanoid Apes can be attributed to it. It had an effect on vegetation all over the world. Man's activities and his impact on flora and fauna fall within its most recent parts. If 3 million well-documented years are so significant, the gaps in our knowledge of the distant past must be equally or more significant, and our conclusions at best should be tentative.

PREPARING THE EARTH FOR AIR-BREATHING LIFE

PETER RICKWOOD

The first rocks of the Earth's crust formed early in the life of the planet when a skin of molten magma cooled and crystallised to form igneous rock. It is likely that a primitive atmosphere escaped from the mantle as the crust cooled. This primaeval atmosphere may have contained most of the nitrogen which comprises the atmosphere now, large quantities of carbon dioxide and hydrogen sulphide as a result of volcanic activity, traces of argon and other gases, water vapour, but no oxygen.

The early land must have been hot while the crust was thin. Volcanic activity, with distortion of the thin crust and related phenomena, suggests that at this time Earth was a violent and awesome place.

Volcanoes erupting on Iceland paint a picture of what it may have been like when Earth was young and its crust was cooling.

THE EARLIEST LIVING ORGANISMS — FROM PROKARYOTES TO EUKARYOTES

The earliest living organisms, Prokaryotes, were microscopic and had outer membranes with a small amount of genetic material simply disseminated throughout the cytoplasm of their single cell (or filament of cells in the case of some Cyanobacteria). They reproduced by simple division into identical cells. Any variation in the characteristics of succeeding generations was largely the result of mutations. When light-harvesting pigment, protochlorophyll, evolved — enabling the organism to produce its own food within its cell — it too was disseminated in the cell contents. Cyanobacteria alive today still retain this primitive arrangement, just like that of their ancestors some 3500 million years ago.

Photosynthesis carried out by green Prokaryote cells produced oxygen as a by-product, and this eventually created the oxygenated atmosphere that characterises our Earth. Thus it was plant life which created a living planet and determined the path that evolution would take.

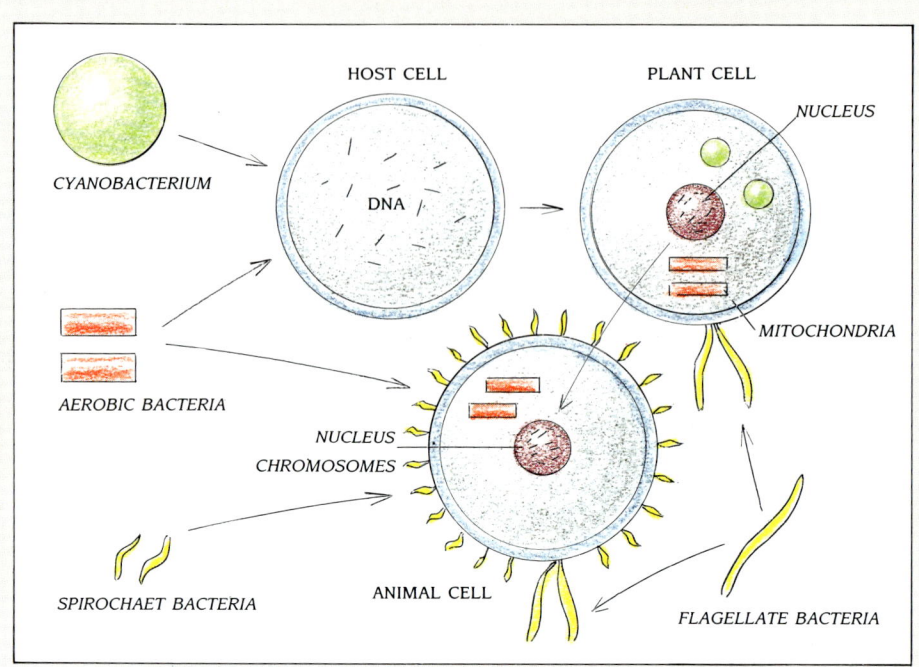

PLANT AND ANIMAL EUKARYOTE CELLS FORMED BY SYMBIOSIS

Living subtidal Stromatolites in Hamelin Pool, Shark Bay, Western Australia. The mushroom-shaped structures are slowly created by Cyanobacteria, and those pictured here are probably several thousands of years old.

The oxygen produced by the earliest green Prokaryotes was initially toxic to many of the simple primaeval organisms with which they co-existed, but over time adaptation to an oxygenated environment was achieved by a range of primitive organisms: saprophytic or parasitic non-green cells (either ingesting dead organisms or other living cells); Bacteria which functioned by using oxygen as an energy source; other Bacteria which had hair-like projections (cilia and flagellae) that propelled them through the water; and the blue-green Cyanobacteria which produced their own food and thus increasingly formed the basis of food chains.

About 1400 million years ago, much more complex cells representing a great evolutionary advance appeared. These Eukaryotes had a cell membrane and within their cell contents were distinct organelles, each enclosed in its own membrane. The most important of these organelles was the nucleus, in which the genetic material was confined and organised into chromosomes. Rod-like organelles, known as mitochondria, were also present in all cells and used oxygen to

produce energy for cell functions. Plant cells had chloroplasts (organelles in which the green pigment was contained). Both plant and animal cells achieved mobility by means of cilia (hair-like projections of the cell membrane) or flagellae (beating tails).

Dr Lynn Margulis of Boston University suggests that the Eukaryote cell was the result of symbiosis between a number of Prokaryote cells. This symbiosis involved a colourless simple cell of proto-amoeba or of proto-fungal type ingesting a Cyanobacterium which became its chloroplast and a number of oxygen-burning Bacteria which became its mitochondria to form an ancestral plant cell. Another organism took over control of the genetic material to produce a nucleus. Mobility was achieved by the attachment of spirochaet or flagellate Bacteria to the cell's outer membrane. The Eukaryote cell which is classified as animal lacks the chloroplasts of plant Eukaryote cells but otherwise acquired its organelles in the same way as did plant cells. There is some support for this theory in that the nature of DNA in the various types of organelles differs.

Living intertidal Stromatolites in Hamelin Pool, Shark Bay.

Fossil Stromatolites, aged about 1600 million years, in the Paradise Creek area, Mt Isa district, Queensland.

The early seas, formed from the condensation of water vapour, would have been warm. How rapidly they became salty, and the concentration of solutes in them, is not known. The concentration of salts in the oceans of the last 1000 million years has remained within the range of tolerance of advanced living cells. Recent evidence suggests that hot springs on mid-ocean ridges regulate the salinity of the ocean. Perhaps there have always been mechanisms, inorganic or organic, which have prevented the concentration of salts from becoming higher than was desirable for living organisms.

Life started in the early oceans. Its origins have been the subject of much speculation. The ammonia from which it was most probably synthesised could have been produced by hot springs related to submarine volcanoes. Bacteria are assumed to have been the first life-forms — the earliest fed on other Bacteria, dead or alive, in the living "soup" in which they swam. Then some acquired the ability to synthesise their own food within the cell. These "autotrophic" forms were a step further up the evolutionary ladder.

Some 3500 million years ago, ancestral Cyanobacteria existed in multitudes. These microscopic organisms had characteristics which place them halfway between the Kingdoms of Bacteria and Plants. They contained ancestral chlorophyll, a green pigment which enabled them to manufacture carbohydrates by photosynthesis. (In this process, carbon dioxide is combined with hydrogen which is obtained by splitting the water molecule, H_2O, using solar energy. The oxygen freed from the water molecule is a by-product.) The activities of the Cyanobacteria, which precipitated lime and trapped silt, resulted in the formation of Stromatolite limestone reefs in shallow-water environments. Stromatolites are the oldest record of life preserved in the Fossil Record. Those found at the North Pole locality in

Early Cambrian fossil Stromatolites at Eurowie Creek in the Georgina Basin of central Australia. The age of the Arrinthunga Formation in which they occur is about 565 million years.

A dolomite outcrop in the Oscar Range, Canning Basin, Western Australia. Dolomite is a rock composed of carbonates of lime and magnesium in equal porportions.

Western Australia, and dated at 3500 million years, are among the most ancient preserved anywhere on Earth.

It was the photosynthetic activities of such organisms and probably untold numbers of others that left no fossil record, over a period of about 2000 million years, which steadily generated oxygen. (There is no better example of life modifying the environment, or of the nexus between the living and the inanimate aspects of the total global environment.) This early life with its green pigment enabling it to photosynthesise (the characteristic that distinguishes Plant life) determined the evolutionary pathway which would be taken by higher life-forms on Earth, and once this step had been initiated there was no turning back. The Plant Kingdom would have been able to evolve without animals but the Animal Kingdom could not have evolved without plants. Earth is the green planet and it was predestined to be so more than 3.5 billion years ago. Some 99.9 per cent of life on Earth today is dependent on plants as a basis of the food chains, yet we still take the Plant Kingdom for granted and fail to realise its essential role as moderator and controller of environmental factors.

At first the oxygen released by the earliest photosynthesisers was used up in the oxidation of many elements and minerals in the seas. As the iron dissolved in sea water was oxidised and precipitated out, ironstone formations were laid down in the shallow epicontinental seas and in troughs along continental margins. Banded ironstone formations worldwide, and the Hamersley and other iron ore deposits in Western Australia, are evidence of the oxidation processes which were occurring between 3800 and 2000 million years ago as oxygen was generated.

When the "oxygen sink" in the oceans had been satisfied, oxidised continental sediments such as red-beds (sedimentary layers coloured by oxidised iron), gypsum and marine evaporites suddenly became abundant, indicating that oxygen had invaded the atmosphere. For oxygen to accumulate in the air and sea it was necessary for carbon to be removed from the system, otherwise oxygen would bond to it and no longer be free. The difference between Earth being a "dead" and a "living" planet is the maintaining of the delicately balanced state where free oxygen is possible. Thus the burial of carbon in organic matter and its transformation into black shales, coal and oil deposits, together with the formation of limestone and dolomite (carbonates), were essential processes involved in creating an atmosphere suitable for breathing.

There were micro-organisms of all types during the evolution of life in the early seas: Bacteria, some of which burned oxygen to obtain their energy; fungal cells which were saprophytic, living on the waste products of other living things; animal cells which ingested other organisms for food; plant cells with green pigment and the ability to photosynthesise; and some cells with intermediate characters. The early forms were all of a primitive cell type, without nuclei or organelles, known as Prokaryotes.

Free oxygen was available in the air from about 1800 million years ago. Presumably it had to reach a critical level before conditions were suitable for more advanced types of cells, the Eukaryotes. These cells are of the type of which multicellular plants and animals are constructed. The genetic material is confined in a nucleus, and there are organelles (small bodies with membranes separating them from the cytoplasm, or cell content). The organelles are of two types: mitochondria, which are rod-like structures that burn up oxygen to supply the energy for cell function in plant and animal cells; and chloroplasts, which contain the green pigment in plants only. The first Eukaryotes probably appeared about 1400 million years ago, when the environment provided the spur to a new burst of evolution — oxygen in sufficient abundance. The origin of the Eukaryotic cell may have been a symbiosis between a number of different Prokaryotes which clubbed together for mutual benefit.

A vast explosion of simple Eukaryote life-forms occurred during an interval of possibly 500 million years after their first appearance. It has been suggested that the activities of green cells during this time led to the depletion of the carbon dioxide in the atmosphere and caused an "Icehouse Effect" (the opposite of the Greenhouse Effect) which could have been a factor contributing to the 300-million-year-long Precambrian ice age that followed. The ice age had peaks at 940, 770 and 620 million years ago and was so intense that the world very nearly froze to death. Even some equatorial regions were under ice sheets.

Such an environmental event of this lengthy duration must have caused massive extinctions of the microscopic life in the seas. The changing and ameliorating conditions as the Earth emerged from the ice age acted as a spur to the next major surge of evolution and ushered in the Age Of Visible Life. Thus most of the categories of Invertebrates were established by the start of Cambrian time about 570 million years ago, and Algae represented the Plant Kingdom. (Oxygen levels may have been not more than 10 per cent of present values, and the ozone layer would have been correspondingly less efficient. Harmful radiation bombarding the Earth was a factor which confined life to the water for the first 150 million years of the Phanerozoic Eon.)

It is fascinating that microscopic plant life manipulated the global environment on such a massive scale throughout the Cryptozoic Eon. Green cells produced an oxygenated world and thus predestined the pathways that subsequent evolution on Earth would follow. The oxygen produced was a waste product of the photosynthetic process. Certainly, the significance and power of the green cells is most dramatically emphasised by the probability that an Icehouse Effect resulted from their using up so much carbon dioxide that the Earth's warm blanket which protects it from the cold of Space became impaired. We must look at our modern world in a much more informed way, for the microscopic algal life in the sea is still an absolutely vital part of the balancing mechanisms which maintain conditions suitable for life. In our modern world about 45 per cent of the carbon dioxide which is removed from the atmosphere is taken up by marine phytoplankton, and an equivalent amount of oxygen is released. Our world has a land flora which through photosynthesis produces oxygen and cycles carbon. The contribution of land-plants is much more generally recognised than that of marine forms. We should be equally or more worried about the sea, for if the day comes when we find we have killed the sea by pollution, acid rain or subjecting it to extra radiation as a result of our damage to the ozone layer, it will be too late and we will be doomed. If the sea dies, the planet dies.

The essential role that micro-organisms of the five Kingdoms — Bacteria, Protista, Fungi, Animal and Plants — play (in addition to that of marine plankton outlined above) in maintaining the balances and keeping conditions on Earth suitable for life remains a little noticed and only partly understood facet of environmental science. They are involved in the chemical processes which keep the proportions of gases in the atmosphere in balance: in the fixing of nitrogen to make it available for living organisms via plants, in keeping the salts in the sea in balance, in removing carbon from the system by the decay of organic matter, in recycling trace elements, and in a host of other functions. In sum, their activities are essential in keeping the balances which allow the planet to remain alive. They have been active in their roles as moderators, controllers, even predestinators, since the dawn of life.

Ore loading at Mt Whaleback.

The East Pit at Mt Whaleback Mine.

BANDED IRONSTONE FORMATION (IRON ORE)

A cut and polished slab of folded banded ironstone from Mt Whaleback Mine, Western Australia.

Tightly folded banded ironstone in a gorge south of Mt Newman, Western Australia. Banded ironstone formations were created when oxygen produced by marine photosynthesisers oxidised the iron dissolved in sea water, before there was an oxygenated atmosphere.

OPPOSITE PAGE:

Red-beds. The "Painted Wall" of Tumblagooda Sandstone is part of a unit more than 1800 metres thick which is believed to have been laid down over a 130-million-year period from the Mid Cambrian through Ordovician and Silurian times.

A deposit of low grade siliceous bauxite. Bauxite is the chief source of aluminium. It is produced by the weathering of igneous rocks under tropical conditions, and the iron oxides in the weathered materials are responsible for the red colouration.

Red-beds coloured by oxidised iron in the Tumblagooda Sandstone sedimentary sequence.

Tumblagooda Sandstone with red-beds.

Laterite at Red Falls, Townsville district, Queensland. Laterites are residual deposits of hydrated iron oxide formed from the weathering of rocks, especially iron-rich basalts, under tropical conditions.

THE EDIACARAN EXPERIMENT

An important Invertebrate fauna which appeared before the "Age Of Visible Life" is remarkable because of its antiquity and also because it is a rare example of the preservation of soft-bodied organisms as fossils. Known as the Ediacaran Fauna, it provides an insight into the evolutionary process.

This shallow-water marine fauna of soft-bodied creatures occurred about 630 million years ago and was possibly widely distributed in low latitudes, as evidenced by the occurrence of isolated members of the fauna in several lands which then lay in the tropics. The most famous example, comprising the most complete fauna, is the fossil assemblage at Ediacara in the Flinders Ranges of South Australia. In a coarse-grained sandstone known as the Pound Quartzite are impressions of organisms which look at first sight like Jellyfish, Seapens (Soft Corals), Annelids (segmented Worms) and primitive Arthropods. It is by no means

Dickinsonia, a broad, flat, bilaterally symmetrical Worm. (Specimen U.NSW. Magn.X 3.1)

certain that the organisms are directly related to living members of the groups to which they appear to belong. Indeed, it seems more likely that they represent a unique experiment in the basic construction of multicellular animals and that they are "experimental" creatures on an evolutionary pathway which was abandoned when the more efficient Metazoan design evolved.

The Jellyfish-like members of the fauna are constructed with concentric structures in the centre of their domed body and radial grooves towards the outside. True Jellyfish, in contrast, have concentric muscles in a ring on the outer edge of their bell with the radial grooves for directing food towards the mouth lying within this ring, towards the centre. So the Ediacaran animals have a reversed structure. The Seapen-like animals also have a different structural pattern from true Soft Corals. The latter have separate branches, often arising from a single stem, allowing free flow of water to the individual polyps which grow on the branches. The Ediacaran examples have a plate-like structure without separate branches and could not have

Madrigania, a Jellyfish. (Specimen AMF. Magn.X 0.9)

functioned in the same way.

The basic architecture of the Ediacaran animals is for light and thin structure with maximum surface-to-weight ratios. They looked like leaves, pancakes or films, sometimes appearing somewhat inflated like small lilos with quilted surfaces. Their adaptive evolution was to obtain size (which gave them more functional surface) without supporting and strengthening mechanisms. This experimental plan was an alternative to the evolution of complex internal organs, which was the more efficient and less size-limiting system adopted by the ancestors of the Metazoa, the multicellular animals characteristic of the Phanerozoic.

Jellyfish. (Specimen AMF. Magn.X 0.9)

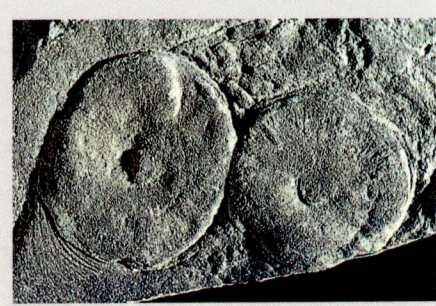

CLIMATE AND THE LIVING EARTH

Climate may be defined as the sum of the prevailing long-term weather conditions in a specified part of the globe. It is determined primarily by latitude but is modified by local conditions such as altitude and the position of the particular region in relation to large landmasses and to the main circulation belts of the atmosphere and the oceans.

The global climate is an engine whose fuel is the heat energy of the Sun and whose parts are land, ocean and atmosphere. Basic circulation patterns in the air and in the sea are the result of the differential heating of the Earth's surface from the Equator to the Poles. The Sun's rays strike the Earth vertically at the Equator and are inclined at increasing angles to the surface towards high latitudes, resulting in unequal transference of heat.

The primary zonation into climatic belts is based on these temperature differences. In the tropics or equatorial belt the hot air rises and flows north and south, spreading the warmth. As it rises, the air cools, and the amount of water it can hold decreases so that rain falls in an equatorial rain belt. While the mass of warm air is travelling north and south, it is cooling gradually, and between 25° and 35° latitudes some of it begins to sink and to flow back towards the Equator, forming the subtropical high pressure zone. As the air sinks it becomes compressed and is able to hold more moisture, so the subtropics have much less rain than the tropics. The rest of the warm upper air continues through the temperate zones towards the high latitudes, cooling as it goes, and on sinking in polar regions it returns towards lower latitudes as a cold surface wind.

This simple picture is complicated by a force generated by the rotation of the Earth, which modifies the basic circulatory pattern. Known as the Coriolis Force, it affects all currents on Earth, in both air and water. Currents are turned clockwise in the Northern Hemisphere and anticlockwise in the Southern Hemisphere. Thus the Coriolis Force converts the surface airflow from the subtropics towards the Equator into trade winds, which blow all year round between 10° and 25–30° latitudes, and similarly the air flowing back from high latitudes towards the Equator is converted into polar easterlies. Between these two bands, within latitudes of 40° and 60°, are bands of westerlies.

A high-altitude wind of great velocity, known as the jet stream, is another factor complicating the simple circulatory plan. It flows in a pattern of large horizontal curves and its pathways affect weather patterns. It is possible that very long-term changes in the wave pattern of the jet stream may have contributed to the ice ages which interrupted the generally warm state of the Earth through geological time.

The seasons also complicate the simple circulation pattern because of the unequal heating of temperate and polar regions throughout the yearly cycle.

The "planetary circulation" depends only on the presence of the Sun, an atmosphere and a spinning Earth, which have been constants throughout time. Thus, throughout the history of the Earth, the primary features of climate have existed. There have always been tropics and subtropics, trade winds, westerlies and a jet stream. There have always been seasons marked

MARY WHITE

Red sky at morning . . . In the Australian context this is not always a "shepherd's warning", not a sailor's, as bushfire smoke in the atmosphere in summer is frequently the culprit. Fire has become a far too constant and devastating fact of life in this arid land since the coming of humans.

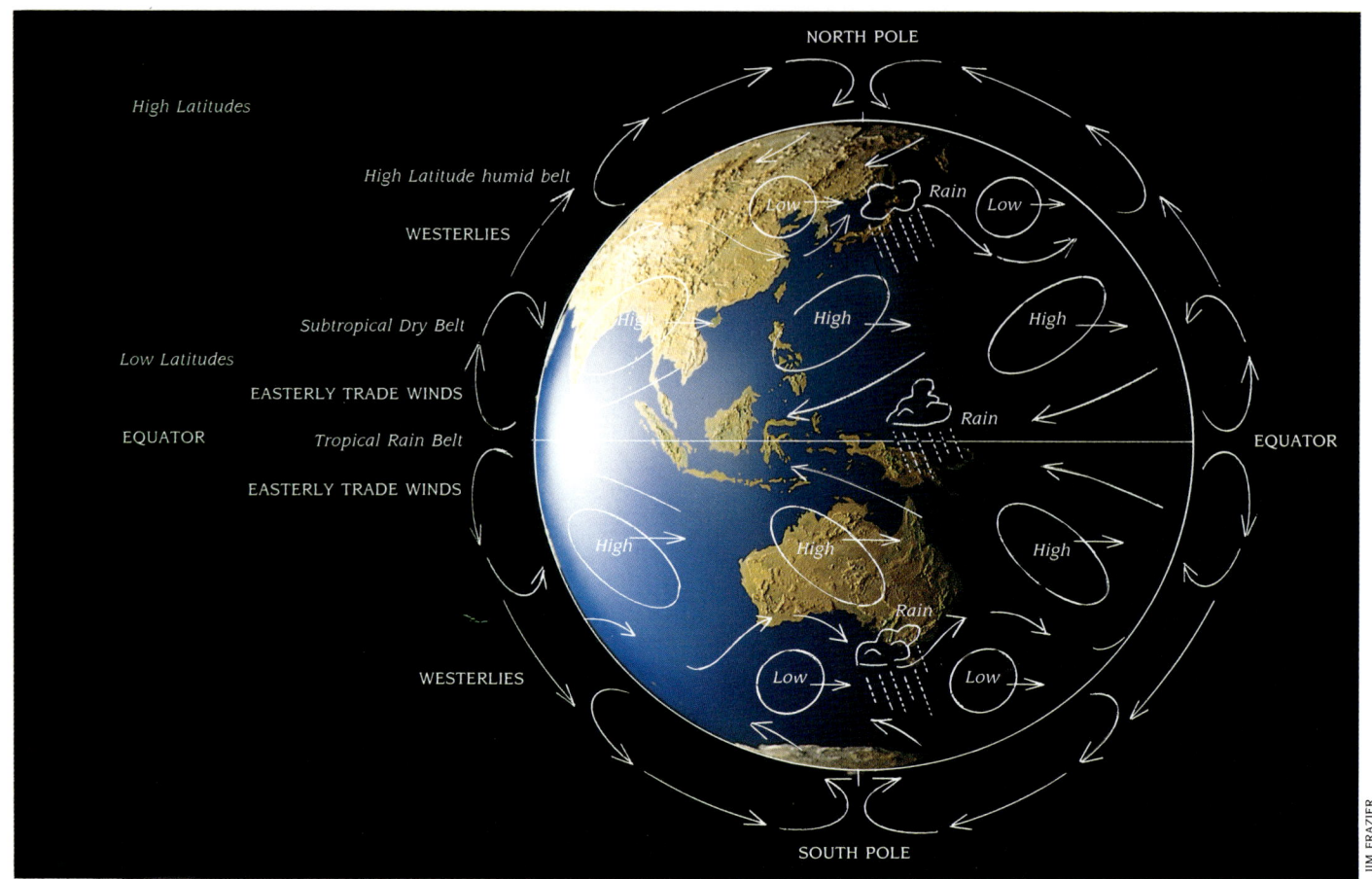

THE PLANETARY CIRCULATION OF THE ATMOSPHERE

The Earth's climate is a solar-powered engine. The Sun's heat warms the land, the air and the water. Over the ocean it causes evaporation, and clouds form which travel and shed their rain elsewhere. The angle at which the Sun's rays strike the Earth alters according to latitude. In high latitudes they strike at a low angle and less heat is transferred than at the Equator where they strike straight downwards. Warm air rises over the Equator and spreads north and south. As it cools on rising, its ability to hold water diminishes and a tropical rain belt results. Between 25° and 35° latitudes, part of the now-cooler air mass sinks and travels back towards the Equator. It is compressed and forms the high pressure systems of the subtropics (which are dry because compressed air has a greater capacity to carry water). The rest of the air mass that rose over the Equator continues towards the Poles, becoming colder as it travels through higher latitudes. Ultimately it sinks in polar regions and travels back as a cold surface wind.

by variations in the length of day and night and by temperature differences. In high latitudes, winter has always brought months of darkness and summer has always had corresponding periods with midnight sun. (However, the overall global temperature and the temperature gradients from Equator to Poles have varied greatly through geological time and the modern connotation of the names for climatic zones can be misleading. The most obvious example is the polar regions, where for much of geological time there was no ice or snow and land was vegetated.)

The planetary circulation patterns of the Earth have been modified by disposition of continents and ocean basins on the surface of the globe. Land heats up and cools down more rapidly than water does. On a daily scale, the land heats up and the warmed air above it rises, causing the wind to blow in from the sea. At night the airflow is reversed because the sea holds its heat for longer. This process, on a seasonal basis, results in tropical monsoons in summer when heated air rising over a hot continent causes moisture-laden air to move in from the sea, dropping its rain on the land.

As with atmospheric circulation, the primary pattern of ocean circulation is determined by temperature differences and by the forces caused by the spinning of the Earth. The arrangement of landmasses and ocean basins on the surface of the globe interferes with the basic pattern of currents and affects climates. When lands are so positioned that they block the flow of warm equatorial waters to high latitudes, the transfer of heat is restricted and the world is cooler than when warm waters can travel freely.

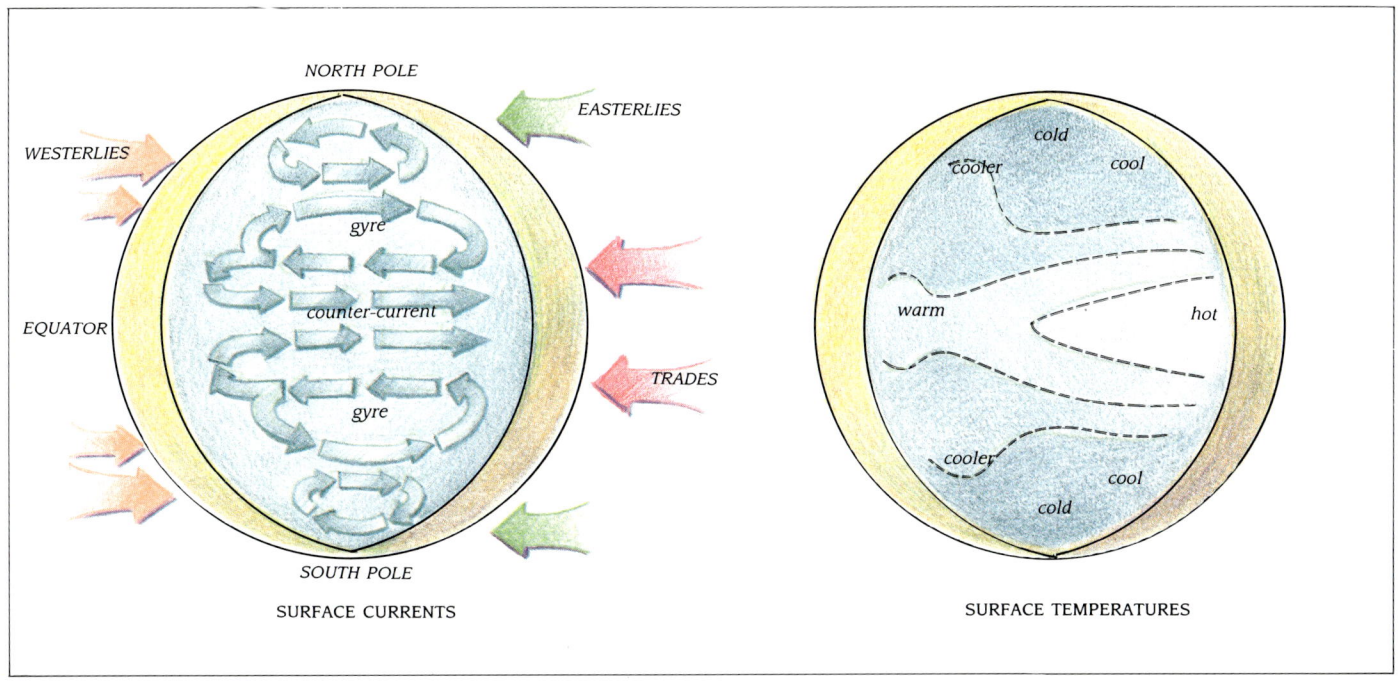

NORTH POLE

WESTERLIES

EASTERLIES

EQUATOR

gyre

counter-current

gyre

TRADES

SOUTH POLE

SURFACE CURRENTS

cold

cooler

cool

warm

hot

cooler

cool

cold

SURFACE TEMPERATURES

PRIMARY CIRCULATION IN A SIMPLE OCEAN

In a simple ocean surrounded by land, surface currents are driven by the planetary winds. The Equatorial Current is driven westwards by trade winds and on encountering the opposite shore it is turned back as a counter-current with two large gyres, one in the Northern Hemisphere and one in the Southern Hemisphere. The surface temperature distribution pattern follows the current pattern, with the warmest temperatures at the end of the equatorial counter-current.

To estimate the climate in a specified area of land at a specific time in the geological past, it is necessary to know how the landmasses were arranged and also where the region in question was situated. By knowing the latitude in which the land lay, and the circulatory pattern of the times, a picture of the climate in that general zone is obtained. Fortunately, our knowledge of the movement of landmasses on the surface of the globe through time is good. Reconstructions on the basis of Plate Tectonics have enabled us to see the changes which have occurred through time, and we are therefore able to assess the general climate. To this general picture must be added the local factors which would also have influenced the region. Topography may have a profound effect, not only in the influence which altitude has on climates, but also because, for example, high mountains can create rain-shadow areas and deserts may exist within geographic zones which are otherwise well-watered. The nature of the land also has a bearing on its climate. Isolated islands have different regimes from pieces of land situated in the middle of supercontinents.

Evidence on latitude, on tectonic movements and on the conditions which prevailed at the time when the rocks of an area were being formed can all be obtained from a study of the rocks themselves. The geological history of an area can be interpreted, and the environment in which the animals and the plants of a specified time lived can be pictured with some degree of accuracy. However, it has to be remembered that a mosaic of microclimates, specialised habitats and different communities existed in the past as in our times. For individual organisms, the interactions between competing species and various other factors have been important in determining whether they survived or became extinct.

READING THE ROCKS TO INTERPRET PAST CLIMATES

It may seem odd that by studying things as lifeless as rocks one can find out what conditions were like for living organisms of the distant past. However, rocks can contain evidence on the latitude at the time of their formation and on the nature of the environments in which they were formed, and, when they contain fossils, the types of living organisms preserved in them indicate the ancient environment in which they once lived. (Part of the romance of fossils is the surprising information they bring of a very different world from that of the present. Marine fossils high up among the peaks of the Himalayas and marine Reptiles in the dead heart of Australia are just two examples that capture the imagination.)

In the same way that fossils can surprise us by telling of a different world in the past, so the information on latitude which is coded into rocks can show that the arrangement of landmasses was very different at times when

Ripple marks in a sedimentary rock, which indicate that the sediments which formed the rock were deposited in the quiet waters of a low-energy environment.

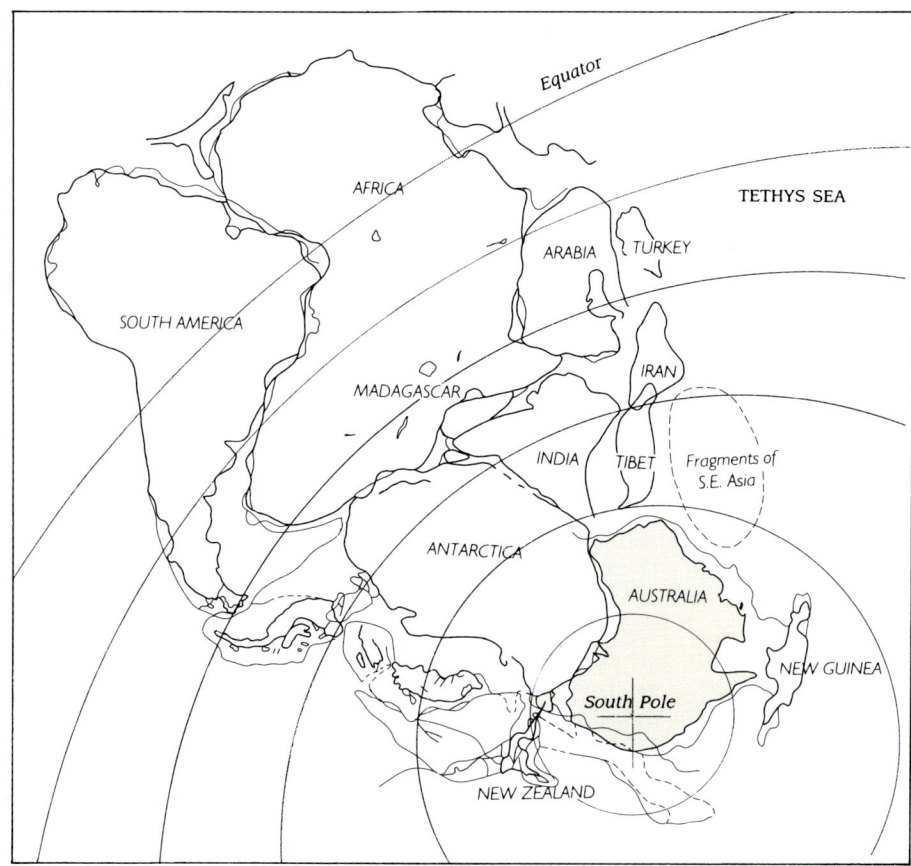

GONDWANA, THE GREAT SOUTHERN SUPERCONTINENT
LATITUDE LINES FOR THE TRIASSIC PERIOD AT 200 MILLION YEARS AGO,
WHEN GONDWANA WAS PART OF PANGAEA

specific rocks were formed. It is the clearly demonstrable evidence about palaeolatitudes which has proved conclusively that continents have moved, that supercontinents have existed and then fragmented, and that the processes of widening ocean basins and drifting continents are continuing.

The Theory Of Continental Drift was propounded primarily to explain the neat fit of the continents' margins when they were brought together, and the observed disjunct distribution patterns in living and fossil plants and animals. The mechanisms by which the arrangement of land and sea on the globe changed were explained only when technological advances enabled the compilation of submarine maps of the ocean basins and also the measurement of palaeomagnetism and magnetic reversals.

Many rocks contain magnetite (an iron ore with magnetic properties), particularly igneous rocks and sedimentary ones made from sediments eroded from igneous rocks. When the rocks were formed, the magnetite crystals aligned themselves with the North–South axis thus coding them with the polarity of the time. Measurement of this "palaeomagnetism" can determine latitude as well as the north-south direction. A picture of the changed positions of bits of the Earth's crust emerges as evidence is accumulated about the polarity and latitude at the time of formation of rocks of a known age in a particular area. It becomes possible to make palaeogeographic maps showing the arrangement of landmasses and oceans at specific times. The climatic zones, dictated by latitude, in which different lands were situated are then evident. The probable circulatory patterns of ocean and atmosphere, as determined by the distribution of land and sea, are revealed by these global reconstructions.

By the start of the Phanerozoic Eon, the lands which were to become today's southern continents and the subcontinent of India, as well as other minor pieces of crust which would later form part of the northern landmasses, were already united. The great supercontinent which they

An Antarctic Beech forest in South America. Similar forests of Nothofagus occur in Tasmania, New Zealand and in the high latitude islands near Antarctica. It was the modern distribution pattern of Nothofagus forests which first suggested to scientists that there might have been a southern supercontinent which included a vegetated Antarctica.

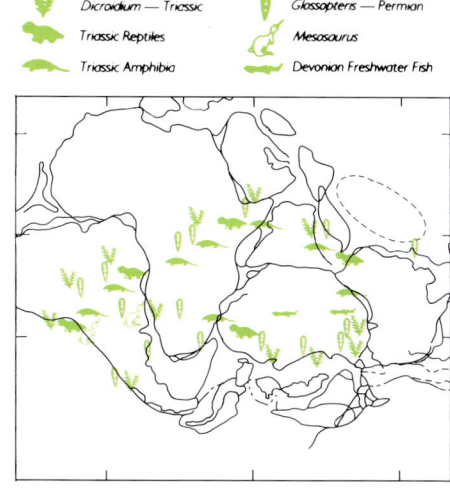

Dicroidium — Triassic   Glossopteris — Permian

Triassic Reptiles   Mesosaurus

Triassic Amphibia   Devonian Freshwater Fish

DISTRIBUTION OF FOSSIL FLORA
AND FAUNA WHICH CONFIRM
THE CONCEPT OF GONDWANA

41

TERRANES

The concept of continents riding on plates and neatly moving apart as oceans widen between them is an over-simplification. The rifting processes are complicated, and bits of continental material can become separated and move independently, later tacking onto other bits or onto the edge of large landmasses. Thus, pieces of crust with quite different geological histories can end up united into one bit of continental crust. Such bits of "foreign" origin are referred to as *terranes*.

New Zealand is a prime example of a modern land comprising a number of different terranes. The complexity of their origins and differences are still the subject of much controversy.

Britain is another example of a land whose parts have different origins and geological histories. Up to 475 million years ago, in the Early Ordovician, England and Wales were part of Gondwana, attached near Morocco. Scotland at that time was attached to the edge of North America. England and Wales then rifted and travelled north as

a long, narrow sliver of land called Avalonia, and became part of Baltica (Scandinavia) which was separated from North America by the Iapetus Sea.

By the late Silurian the Northern Hemisphere had four landmasses: North America (plus Scotland); Baltica (plus England and Wales); and two others. With the formation of Pangaea at the end of the Palaeozoic Era all these lands were amalgamated, and the closing of the Iapetus Sea brought England, Scotland and Wales together. When Pangaea split up again in the Mesozoic Era and the North Atlantic opened, Great Britain went with the Europe–Asia segment and moved to its present position.

TERRANES OF SOUTH ISLAND, NEW ZEALAND

Median Tectonic Line

Karamea / Golden Bay / Drumduan / Brook Street-Murihiku / Dun Mountain-Maitai / Caples / Older Torlesse / Younger Torlesse

Kakahu / Akatarawa

comprised straddled the Equator at the start of Cambrian times and would later swing round into the Southern Hemisphere to become the Great Southern Land, now known as Gondwana.

Landmasses of the Northern Hemisphere were not united at the start of the Phanerozoic but were coming towards each other and would unite and bond on to West Gondwana to form the super-supercontinent Pangaea in the Early Mesozoic, about 200 million years ago. Pangaea, the "one Earth", would stretch from Pole to Pole down one side of the globe, and the mighty ocean, Panthalassa, would wrap around it. An embayment of the "one ocean", the Tethys Sea, would partly separate the northern section of Pangaea, the supercontinent Laurasia, from the southern Gondwana.

Pangaea existed for about 50 million years before it began to fragment and the continents started to move towards their present positions on the globe. The process is still incomplete, and in the case of Australia the movement continues northwards at about 7 centimetres a year.

The separation of continents was achieved by a process of rifting followed by the generation of new seafloor, and it took place at different times between different parts of the landmasses. As the arrangements of land and sea altered progressively, climates were affected. Migration of plants and animals was possible when lands were linked or close, and it became increasingly difficult as distances between them increased.

Other environmental evidence, apart from that which relates to latitude, is available from the study of rocks and regional geology. The sedimentary rocks of any region contain a record of the prevailing conditions at the time of their formation. Local topography influenced both the rate of deposition and the character of the sediments which accumulated to form them. The types of sediment found indicate whether the river systems which deposited them were high-energy or low-energy systems. Hence, they give clues as to the rainfall and the gradient of catchments. Rock sequences formed from sediments deposited in quiet lakes or spread by active streams or blown by wind have different characteristics. Glacial sediments indicate ice age conditions, and grooves and striations etched into the surface of rocks indicate the passage of glaciers or an ice sheet over a region. Volcanic ash or debris incorporated into sedimentary rocks tells of active volcanoes in the region while the rocks were being formed.

Certain rock types indicate specific climatic conditions. The presence of evaporites, produced by the evaporation of saline waters and the crystallisation of salt, gypsum and other minerals, implies aridity. Some red-beds, coloured by the oxidation of iron, are the product of alternating arid and wet climatic conditions. Blown sand or dunes indicate desert conditions.

High water-tables were required for the formation of the peat and coal swamps which produced lignite and coal, and the natural gas often associated with them. The temperature regimes under which coal floras flourished are known to have been widely divergent, from tropical to cold-temperate, so no estimate of climate can be made on the presence of coal alone.

Cyclic bedding, in the Pillara Formation of Devonian age, in the Canning Basin of Western Australia. Alternation of fine-grained and coarse-grained sedimentary layers reflects alternating cycles of quiet and active deposition.

MURRAY JOHNSTON

CLASSIFICATION OF ROCKS

There are three main classes of rock: igneous, sedimentary and metamorphic.

The crust of the Earth — the first rock — was formed by the cooling of a thin skin of molten magma. Such rock, and that produced by the activity of volcanoes and the escape of molten magma through cracks and fissures in the mantle, is known as igneous rock. The word is derived from the Latin *ignis* meaning "a fire", indicating the super-heated and molten origins of the different rock types formed in this way. Lavas are igneous rocks which have gushed from volcanic vents; sills are flat-lying sheets of volcanic rock which spread horizontally like a flood across a surface or between layers of pre-existing rocks; dykes are more or less vertical sheets of rock formed by the crystallisation of magma which has forced its way upwards through a cleft (a fault or a line of weakness) inclined at a high angle to the layers of pre-existing rocks; and plutonic rocks are major, and often immense, intrusions which cooled far below the surface of the Earth to form granite batholiths and the like.

Erosion of the original rocks of the Earth's crust and all subsequent rocks has produced sediments which in time became compacted and turned into sedimentary rocks. Sedimentary rocks have themselves been eroded into sediments and rebuilt into new rocks.

Granite tors in the Mt Isa district, Queensland, are dramatic features of the landscape. Granite is a coarse-grained igneous rock consisting mainly of quartz, feldspar and mica, with various accessory minerals. It occurs in intrusive bodies formed from molten magma which crystallised at depth in the Earth's crust. Such bodies of rock are known as plutonic intrusions.

The Hawkesbury Sandstone of the Sydney Basin is a sedimentary rock which weathers into fantastic shapes and patterns.

Sedimentary rocks and unconsolidated sediments cover most of the ocean floor and about three-quarters of the continental land area. They form layers which are seldom more than a few kilometres thick upon the mainly igneous and metamorphic crust. Where there has been major subsidence, sedimentary accumulations can reach a depth of 20 to 30 kilometres. Sedimentary rocks are classified according to the coarseness of their texture and the source and composition of their sediments. Conglomerates are consolidated pebbles, rocks and sands; sandstones may be coarse-grained or fine-grained; silts and mudstones are very fine-grained; shales are fine-grained and distinctly layered (laminated). The sedimentary classification is generally extended to include residual, organic and solution deposits — carbonates, limestones and oozes — produced by the activities of living organisms or from their shells or skeletons.

Metamorphic rocks are altered sedimentary or igneous rocks which have undergone change in respect of their texture or chemical composition through being subjected to high temperature and/or pressure. Rocks may be baked and changed by new igneous intrusions, or pressure and sheering stresses can cause change in structure.

GLACIERS AS AGENTS OF EROSION

Glaciers are rivers of ice. They erode the land over which they travel, and carry and spread sediments in the same way as do rivers of water.

Ice sheets spreading from the Poles to lower latitudes, and glaciers moving from high ground to lower areas, wear down the land surfaces over which they pass. The moving ice sculptures the mountainous landscapes and deposits glacial sediment on the lower regions.

The great abrasive power of glaciers is due to the rocks, grit and sand frozen into their bases, which grind into the land surfaces under the weight of the ice. Rock is ground into "flour" and all grades of sediment larger than that. Rivers fed by glaciers are "milky" with the rock flour held in suspension in their waters.

Glaciers polish the surfaces over which they move, and scour deeply into the bedrock. "Striated pavements", where rocks imbedded in the ice have gouged deep grooves in the bedrock, are a feature of landscapes which have been glaciated. The striations left on the rocks show the direction of the glaciers' passage. Large boulders, too, can be carried by glaciers, either frozen into their basal layers or perched on their surface (if they have fallen from surrounding peaks onto the moving ice).

When a glacier enters the sea and calves off icebergs and sea-ice, the sediments carried in its basal ice can be spread over large areas of the sea bottom. Boulders carried on the surface are occasionally rafted away on ice flows to become "dropstones", which, after having travelled some distance on the ice, find their way into non-glacial sedimentary rocks. Glaciers which terminate on land build up moraines of grooved and striated pebbles and rocks at their snouts and along their margins. Glacial till, a rock type formed from such deposits, is characteristically composed of unsorted rock fragments of different sizes and of sands of all grades.

Braid-plains where rivers twist and join characterise areas of flat land in front of glaciers. Here sediments are sorted by the changing flow of the rivers in winter and summer seasons.

Very fine-grained glacial sediments (rock flour) may be blown into dunes by the wind, forming "loess" deposits. Loess covers large areas of land affected by the last (Pleistocene) ice age in Europe, Britain, North America and Asia. It also occurs in New Zealand.

The Battye Glacier, Beaver Lake district, Antarctica.

MAGNETISM OF THE EARTH AND PALAEOMAGNETISM IN ROCKS

Compasses whose needles point to the North Magnetic Pole are familiar aids to navigation. They function because the spinning Earth acts like a giant dynamo and generates a magnetic field which is strongly directional. "Polarity" is the term used for this directional effect. But the needle of a compass is affected by vertical as well as horizontal magnetic forces, and the vertical effect varies depending on latitude. A suspended magnetised needle, free to pivot vertically, will align itself in the direction of the Earth's magnetic field. At the Equator the compass needle lies essentially flat. Moving towards higher latitudes, the angle of the needle to the horizontal increases until it points straight down over the Poles. The angle (or inclination) of magnetism can be measured accurately and used to determine latitude.

Before Man-made compasses, lodestones (rocks rich in magnetite) were used by early navigators. The iron-rich magnetite, which has natural magnetic properties, occurs abundantly in igneous rocks such as basalts, dolerites and granites (which have been formed by the crystallisation of magma from the molten core of the Earth). It occurs also in sedimentary rocks which

have added the ash and debris of volcanic eruptions to their non-igneous sediments, and abundantly in sedimentary rocks comprising sediments eroded from igneous rocks.

In the case of igneous rocks which have erupted at super-heated temperatures, the molten magma has to cool below 575° Centigrade (the Curie Point) before the magnetite attains its magnetic properties. Each crystal of magnetite then aligns itself in the direction of the Earth's magnetic field at the time of cooling. Provided the rock has not been reheated since its formation by the later intrusion of more molten volcanic rock, it is possible to measure the original magnetic direction and inclination. The rock is, as it were, coded with this information at the time of its formation. Magnetic particles contained in sediments which later form sedimentary rocks also align themselves with the polarity of the times and by measuring the angle of inclination the latitude at the time of deposition can be determined.

Fossil magnetism — palaeo-magnetism — in rocks enables the plotting of the movement of landmasses.

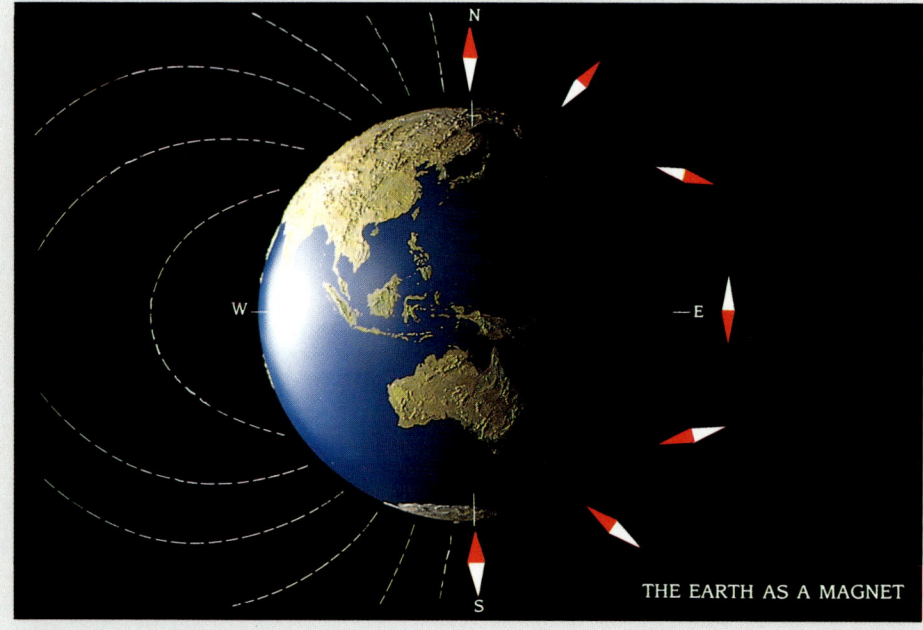

THE EARTH AS A MAGNET

THE THEORY OF CONTINENTAL DRIFT

Early geographers noticed that some continental coastlines, like the immediately obvious opposing margins of Africa and South America, could be fitted together neatly like a jigsaw puzzle. Their observations led them to wonder if there might have been different arrangements of land and sea during the distant past, with some of the continents united into supercontinents. The Theory Of Continental Drift evolved from the seeds of such an idea when scientific evidence on the distribution of plants and animals, living and fossil, and on geology progressively added weight to the concept.

Sir James Hooker, naturalist on James Ross's voyages of discovery in the ships *Erebus* and *Terror* between 1838 and 1843, made a significant contribution to the concept. He recognised the essential unity of the circum-Antarctic floras with their southern plant families when he collected and classified plants in the region.

In 1857 W.L. Green first formulated the actual theory, but he was ahead of his time and of substantiating evidence, and the scientific community of the day virtually did not take any notice of his work. However, as knowledge of living plants increased during the late 19th Century and classification of species proceeded, distribution patterns emerged which could not be explained in the context of widely separated landmasses.

Another 55 years were to elapse before Alfred Wegener published his concept of the changing face of the globe in 1912. Then he, and not W.L. Green, was credited as being the "father" of Continental Drift. Shortly afterwards, the eminent South African geologist Alex L. duToit put drifting continents into the realm of scientific respectability with his study of rock formations and stratigraphy of southern lands. He was able to show the same rocks on opposing margins of now separate continents and similar rock sequences within them.

PLATE TECTONICS — EXPLAINING HOW CONTINENTAL DRIFT WORKS

General acceptance of the Theory Of Continental Drift was not reached until the 1960s due to the fact that until then no satisfactory explanation could be found for the mechanics involved in moving bits of the Earth's crust about. However, with technological advances which made mapping of the ocean floor possible, a new age of geological understanding was born through the new Theory Of Plate Tectonics.

When the underwater topography of the ocean floor was disclosed, it was found that mountain chains lie near the middle of ocean basins and enormously deep trenches near the margins. About 80,000 kilometres of high mountain ranges lie under the sea in the main mid-ocean ridges, and the relief is often more dramatic than that in the Himalayas. The Marianas Trench, for example, is on a similar scale to the Himalayas, but deeper than Mt Everest is high. The significance of the submarine mountain ranges and trenches was realised by Professor Harry Hess, who in 1962 suggested that there might be movement of ocean floor in two directions away from mid-ocean ridges. This "seafloor spreading", he proposed, would move continents apart by increasing the width of ocean basins. Hess's concept was based on the theory that molten rock from the Earth's mantle came to the surface at positions marked by the ridges, then spread laterally at right angles to their long axis and formed new crust. To compensate for the newly formed crust, old oceanic crust was subducted in the trenches.

In 1963 two geomorphologists, Vine and Matthews, reasoned that if the seafloor spreading was occurring it should be possible to prove, because new advances in technology enabled the measuring of palaeomagnetism in rocks and the detection of magnetic field reversals. They argued that the igneous rock produced on the ridges should have been coded with a record of the Earth's magnetic field at the time of formation. In addition, the periodic reversals in the magnetic field which are known to have occurred throughout geological time should be detectable.

Magnetometer surveys subsequently were carried out along mid-ocean ridges and revealed mirror-image patterns of reversals on either side of the ridges, just as predicted. Dating showed that the reversals were episodic, having occurred on average at about 450,000-year intervals. From this information it was possible to calculate the rate of spreading. It was found that the Atlantic Ocean is widening by 3 to 5 centimetres a year. While this distance at first seems small, it represents 20 metres since the discovery of America by Columbus. Dating of seafloor globally has shown that none is older than 200 million years, indicating that older oceanic crust indeed has been subducted. Thus, the theory that seafloor spreading was the force moving continents about was tested and proved.

While most rifting and spreading takes place on the mid-ocean ridges under the sea, rift zones also extend onto some continental masses. Rift valleys are present in eastern Africa, in the Baja California region of North America, and in Iceland — where the Mid-Atlantic Ridge is steadily splitting the land apart.

The partial understanding of the mechanisms involved in the movement of landmasses by the evolution of ocean basins between them has had a profound effect on Earth Science. Subsequently the Plate Tectonic model for rearrangement of land and sea has been refined and accepted.

It is now understood that the crust of the Earth consists of a number of plates, which are rigid and practically undistorted slabs of lithosphere on which the continents ride. The border regions between adjacent plates are "active zones" where most major earthquakes, volcanic activity and mountain-building occur as a result of interaction between the plates.

The number and size of plates comprising the Earth's crust has changed through the ages. The Plate Tectonic model applies only to the last 1500 million years of geological time. Before that, during the early history of the Earth, it appears there may have been many small landmasses moving about and colliding, joining and separating. In that "microtectonic" phase lands were probably small enough not to interrupt the global flow of warm equatorial waters, the thinner crust probably meant warmer surface temperatures, and climates were probably torrid. By about 1500 million years ago the volume of continental crust may have been about three-quarters of its present size.

The movement of plates and the changing disposition of land and sea are well documented for the 570 million years of the Phanerozoic Eon, helping us to assess the climates and conditions under which plants and animals lived in each geological Period. Plates have not always been the same as they are today, either in size or number. Between 125 and 45 million years ago, India was on a separate plate and moved northwards away from Antarctica and Australia. When it was firmly pressed against Asia, spreading ceased on the divergent margin between the Indian Plate and the Australian Plate and the two coalesced to form the modern Australia–India Plate. It has a consuming margin on the north and the east, and an extending margin on the south and the west. Australia and Antarctica shared a plate until rifting between them from possibly 100 million years ago gave them separate plates. Australia's movement away from Antarctica in the last 45 million years has caused India, on the leading edge of the Australia–India Plate, to bore into Asia, pushing up the Himalayas.

BRUNO JEAN GRASSWILL

ICELAND

Reykjanes Ridge

NORTH AMERICAN PLATE

EURASIAN PLATE

IRANIAN PLATE

ARABIAN PLATE

AFRICAN PLATE

Mid-Atlantic Ridge

SOUTH AMERICAN PLATE

Mid-Indian Ridge

South-west Indian Ridge

FAULKLAND TRANSFORM FAULT

SCOTIA PLATE

▲ volcanoes

hot spots

MAJOR PLATES, SPREADING RIDGES AND TRENCHES OF THE EARTH'S CRUST

NORTH AMERICAN PLATE

Aleutian Trench

Japan Trench

CHINA PLATE

Marianas Trench

PHILIPPINE PLATE

Hawaiian Ridge

PACIFIC PLATE

SAN ANDREAS FAULT

New Guinea Trench

INDO-AUSTRALIAN PLATE

AUSTRALIA

Peru–Chile Trench

NAZCA PLATE

Tonga–Kermadek Trench

Lord Howe Rise

NEW ZEALAND

East Pacific Rise

Antarctic–Pacific Ridge

ANTARCTIC PLATE

THE ICELAND RIFT VALLEY

Iceland is presently being split in two by a rift valley located where the Mid-Atlantic ridge rises above the sea. The rift is a dramatic feature of this far northern land. Volcanoes are active and there are earthquakes.

Iceland lies at about 65° North, and an interesting comparison with its latitude is Australia's location in corresponding southern latitudes from Late Carboniferous to Early Tertiary times, particularly just after the ice age of the Late Carboniferous to the Early Permian. Iceland now, after the Pleistocene ice age, and Australia-in-Gondwana in that former time probably had similar climates.

The rift at Thingrellir where the first parliament in history was held. Here on the banks of the Oxara River, about 40 kilometres from Reykjavik, the walls of basalt formed an amphitheatre with acoustics that enabled all people in the assembled crowd to hear what was being said. In the ceremony in the year 930 AD the foundations of the Icelandic Republic were laid when a national code of law was adopted and a General Assembly was established.

PETER RICKWOOD

PETER RICKWOOD

PETER RICKWOOD

Fiery lava flows across a black basalt plain newly formed by the active volcano Krafla which erupted in 1980.

The ridge of basalt on one side of the rift, and a volcano in the background.

The rift which is splitting Iceland apart.

PETER RICKWOOD

SEDIMENTARY BASINS OF AUSTRALIA

The eastern half of Australia has a vertical succession of basins of different ages. The most recent, formed in the Late Mesozoic, are the Eromanga Basin and the Surat Basin. They overlie older basins which had been formed during the Early Mesozoic and the Late Palaeozoic (shown on the map in dashed outline). These older areas of subsidence are the Arckaringa, Pedirka, Cooper, Galilee and Bowen basins. Under them are even older basins formed during the Early and the Mid Palaeozoic; they are the Drummond, Adavale, Darling and Arrowie basins (outlined on the map with a dotted line). Sedimentary rock sequences were laid down in all these basins of different ages.

The basins were downwarps in the crust into which rivers drained, carrying sediments derived from the erosion of local land surfaces and from river catchments which may have been quite distant. Carbonates, which form limestones, accumulated in some basins under suitable conditions as a result of the biological activities of living organisms in the waters of the basins themselves. The weight of accumulated sediment caused further sagging of the crust, deepening the basins and allowing further layers of sediment to accumulate. Layers, or strata, of sediment were compacted and converted over time into sedimentary rock sequences.

Some basins owe their origin to faulting and rifting (cracking and tearing of the crust) during tectonic movements. Whole blocks subsided while others were elevated, and drainage into lowered areas started the accumulation of sediments. Most fossil localities are in sedimentary basins, and many have been found during the systematic geological mapping of the major basins in the course of exploration for coal, oil and natural gas.

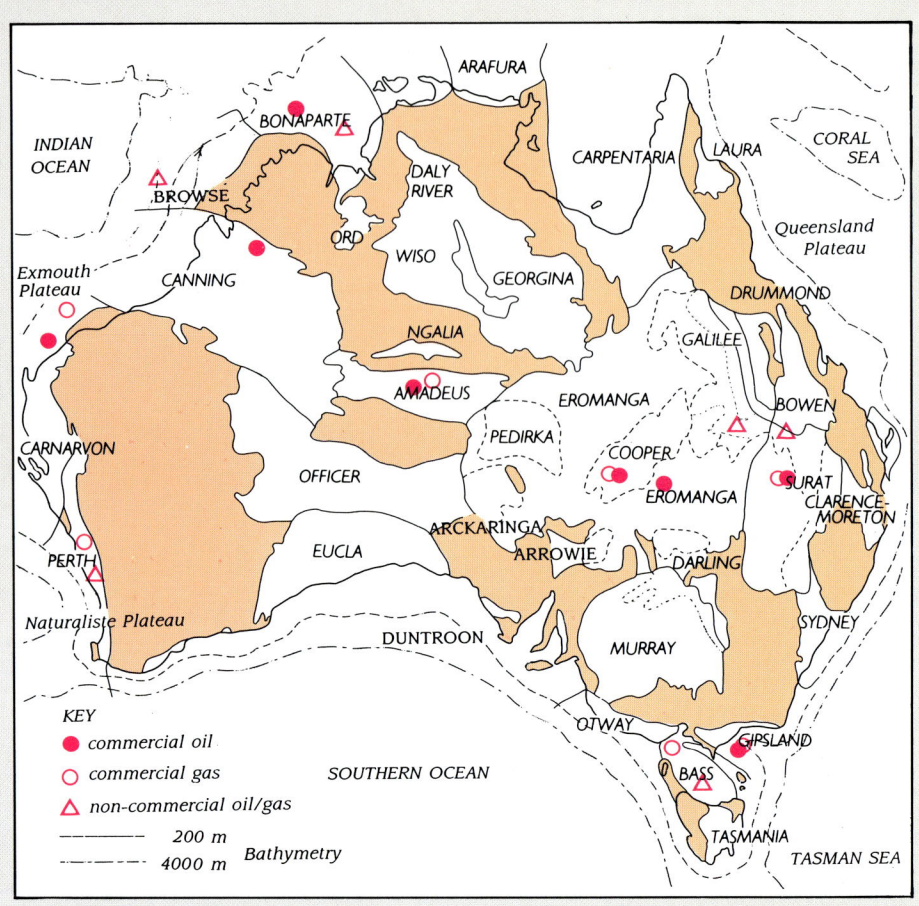

After Evans 1988

SEDIMENTARY BASINS OF AUSTRALIA
SHOWING THEIR ACTUAL AND POTENTIAL CAPACITY TO PRODUCE HYDROCARBONS

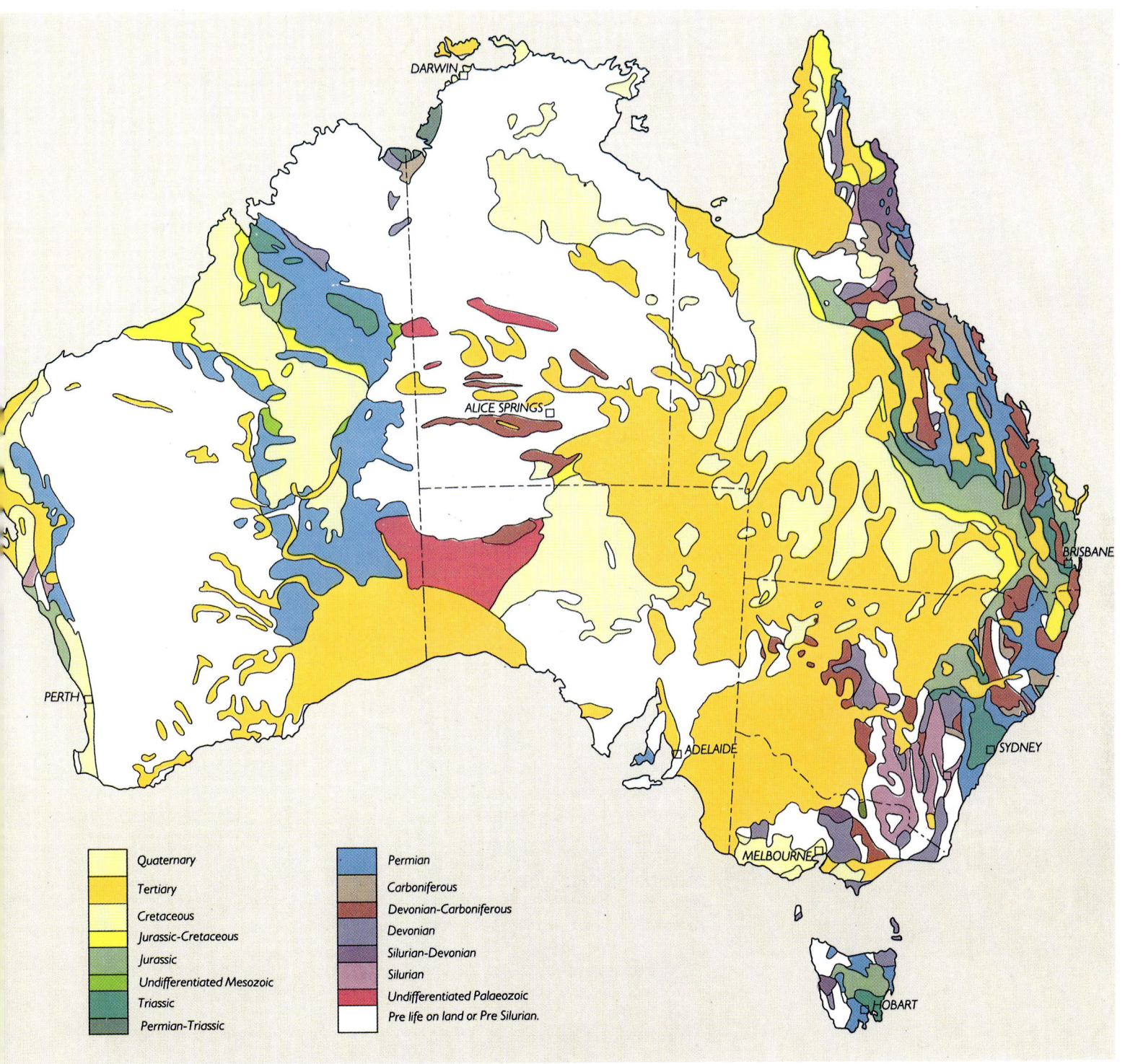

SIMPLIFIED GEOLOGICAL MAP OF AUSTRALIA

Quaternary

Tertiary

Cretaceous

Jurassic-Cretaceous

Jurassic

Undifferentiated Mesozoic

Triassic

Permian-Triassic

Permian

Carboniferous

Devonian-Carboniferous

Devonian

Silurian-Devonian

Silurian

Undifferentiated Palaeozoic

Pre life on land or Pre Silurian.

FOSSILS AS A KEY TO INTERPRETING ANCIENT ENVIRONMENTS

An assemblage of Early Devonian sea creatures, about 400 million years old, fossilised in the mud of a seafloor on which they lived. Gastropod Snail shells, spherical and stick-like Corals (Heliolites and Cladophora) look like marine animals of the present day. (Specimen MQU. Magn.X1.8)

A fossil Fish from the Talbragar Fish Beds, New South Wales. This fossil locality is renowned for its beautifully preserved plant fossils and Fish which lived about 175 million years ago in the Jurassic Period. (Specimen AMF. Magn.X1.8)

The Fossil Record which is preserved in rocks of the Earth's crust contains a sample of the flora and fauna which were alive at the time when the rocks were formed. However, the way in which the sample came to be preserved, as well as the general limitations of this record, must be understood when interpretations of environment are made on the basis of the presence or absence of species of plants or animals.

Most fossils of land-living organisms occur in sedimentary rocks which were made from the compacted sediments that accumulated in basins or hollows in the Earth's crust into which drainage systems fed. Fossiliferous rocks are rare considering the great abundance of sedimentary rocks — special conditions of sedimentation and a number of other factors have to be present before plant fragments or other organic matter can be preserved.

The organisms which appear in the Fossil Record have been selected from the flora and fauna of the times in a biased fashion because of the process by which the sediments themselves accumulated. Only organisms living close to or in the deposition site or river catchments have much chance of appearing in the record at all, and the chances of a species making it into the record unless it was very common and abundant are small. A further limitation is the fact that usually only the more robust parts of plants and animals are likely to survive as fossils. Delicate tissues are very seldom preserved.

Sedimentary rocks formed from deposits in shallow-water marine environments where Invertebrates with hard exoskeletons and shells are abundant are often richly fossiliferous. They probably show a high proportion of the species with resistant parts alive at the time the rock was formed, but lack all soft-bodied and delicate creatures whose tissues were unsuited to preservation. Fish, when they evolved, had a fair chance of appearing as fossils, having bones, teeth and scales suitable for fossilisation. Amphibians obviously had more chance of appearing in the Fossil Record than did Reptiles which lived in dry places, because they spent some time in the water and their larval stages were also in the right place to be incorporated in sediments.

The Fossil Record is at best incomplete and fragmentary. It cannot be used on its own to deduce what conditions were like in the distant past, but in conjunction with other evidence it can make a considerable contribution.

When fossils are used as a key to interpreting the environment of past ages, a basic assumption is often made which may not always be justified. We assume that the environmental requirements of plants and animals of the past were similar to those of their descendants which are alive today. When there is no evidence to the contrary after all the known facts about the environment of a specified time are assessed, such a subjective approach is acceptable because there is no alternative.

In general, in the past as in the present, there is a greater diversity of living things in tropical regions, with gradation to lesser diversity toward high latitudes. In fact, more than 90 per cent of the world's living species

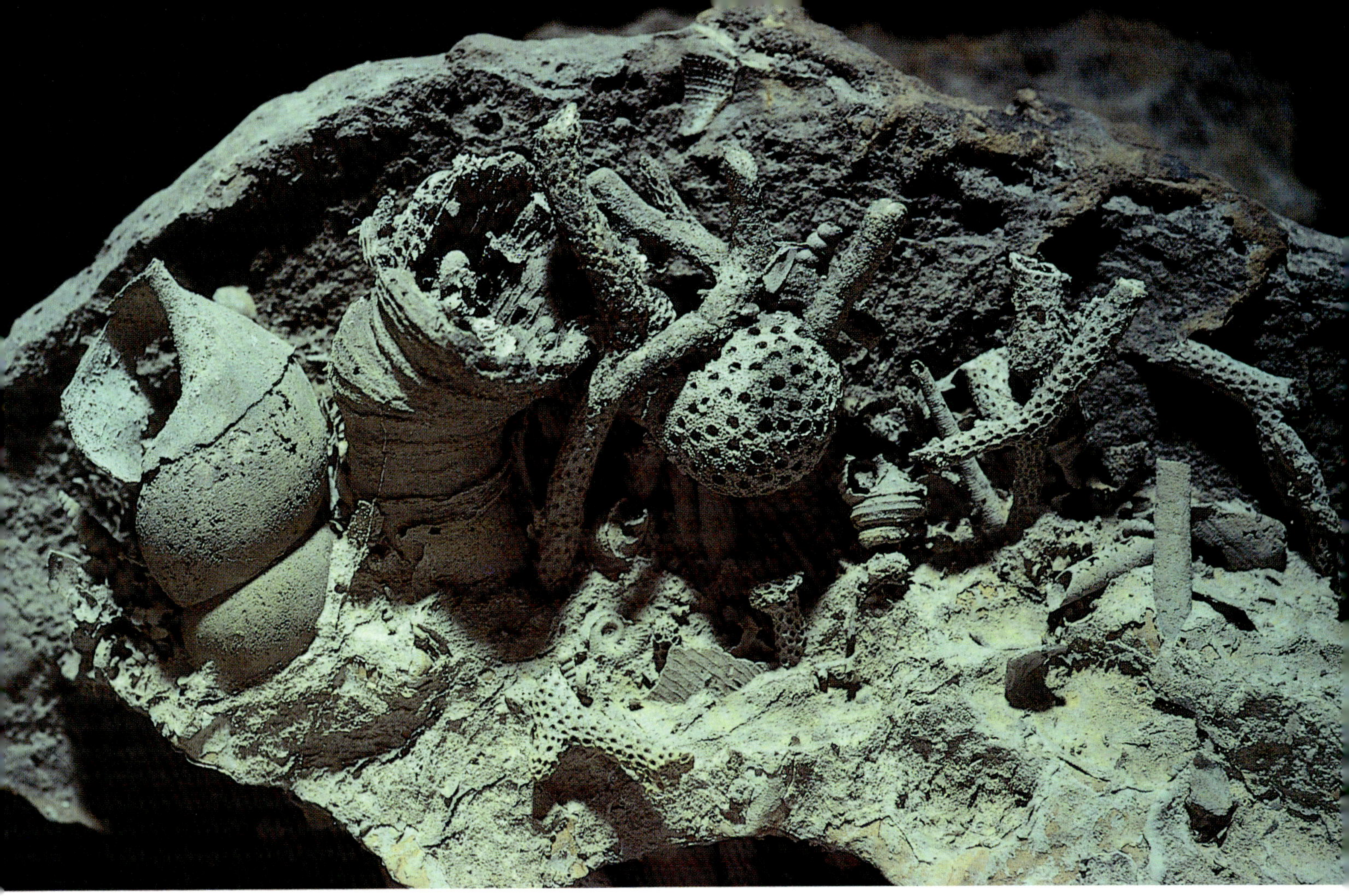

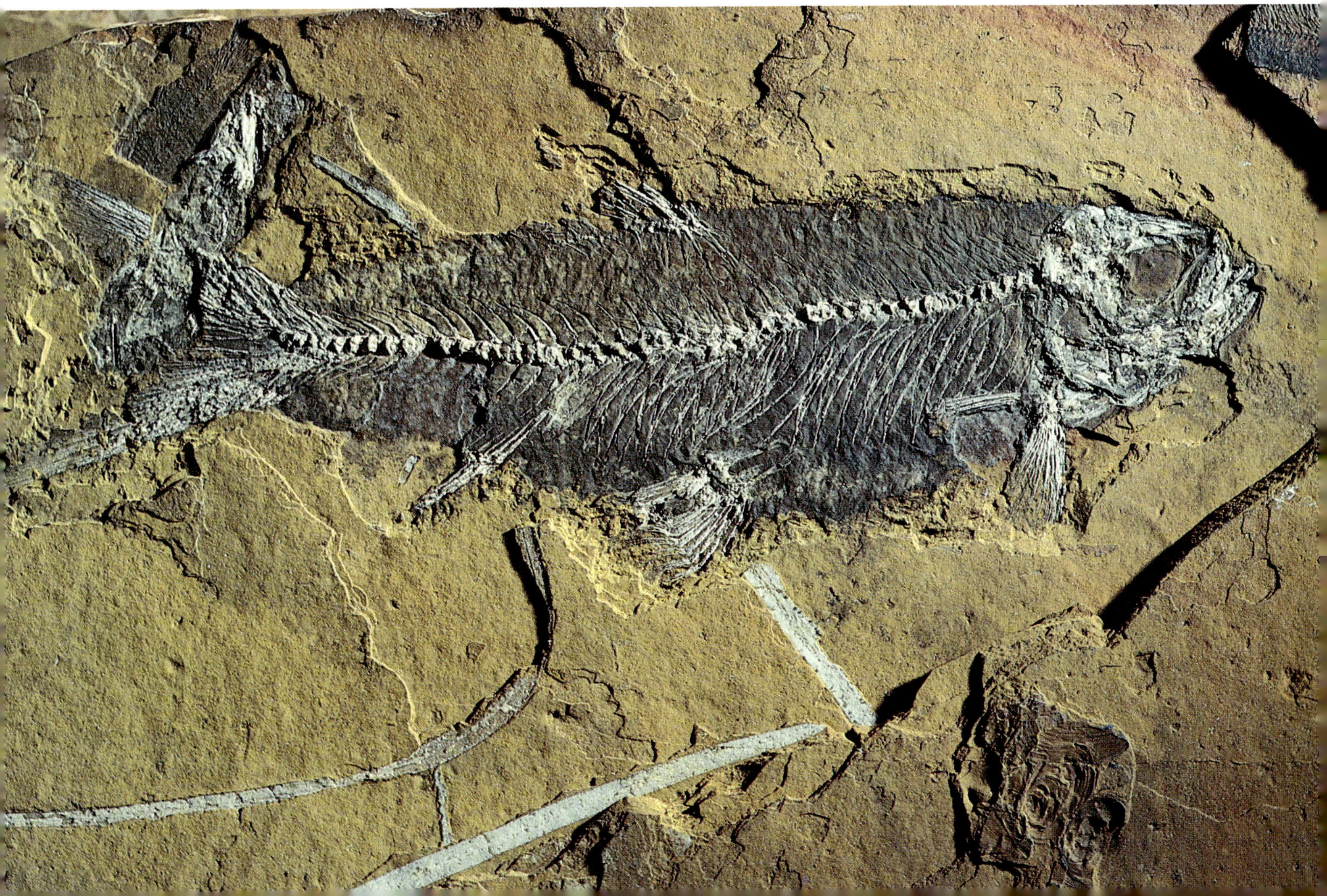

Mid Jurassic Ammonites from the Newmarracarra Limestone at Geraldton in Western Australia. (Specimen WAM. Magn.X 2.5)

occur in tropical Rainforests. Thus a fossil assemblage showing a great number and diversity of species may indicate that the community lived under warm to hot climatic conditions. Conversely, very impoverished assemblages may indicate high latitudes (or ice age situations).

Deposits comprising plankton shells and tests which accumulated on the ocean floor can be indicators of the water temperature at the time when the plankton were living. Calcareous deep-sea oozes indicate warm waters, while siliceous oozes indicate colder waters. Some Foraminifera (small plankton with coiled shells) produced shells which coiled to the right in warm water and to the left in cold water.

It has been shown that Coral reefs and reef limestones had a tropical and subtropical distribution pattern in the geological past, just as they do today. Fossil reefs can be taken to indicate that the regions in which they are found today were in low latitudes at the time when the Coral was alive.

From the Plant Fossil Record it is possible to make certain deductions about the environment in which some plants lived. Modifications to prevent water loss — xeromorphic adaptations — indicate that the plants grew under conditions which were seasonally arid. (Reduction of leaf surface, thick cuticles and other features of drought-resistant plants are familiar to Australians because sclerophylly is a characteristic of many of the country's modern plants.) In the case of the Flowering Plants, which have a fossil history of only about 120 million years, broad leaves with entire margins characterise Rainforest trees, while small leaves with variously cut margins are indicators of cold or alpine regions.

FORAMINIFERA

The Foraminiferids are very small organisms of the Kingdom Protista. Among the first protozoans to develop hard skeletons in Cambrian times, they have left a fossil record from 550 million years ago up to today. Their shells are usually composed of a series of chambers which were added as the organism grew, with the last and usually the largest open to the exterior. Most shells are perforated by small pores through which the living protoplasm of the organism protruded. The earlier types built their shells from pre-existing grains of sediment, but later forms made calcareous (lime-rich) shells. Since Devonian times the multitudes of tiny shells have been a major constituent of limestones.

Foraminiferids were benthic (bottom dwelling) until the Mid Jurassic, and since then they have been planktonic (free floating) and abundant. "*Globigerina* ooze" (Cretaceous to Recent) covers large areas of the ocean floor; indeed, about a quarter of the surface of the globe is blanketed with deposits formed from the minute shells of these organisms.

JOHN CLEASBY AND GUNTHER BISCHOFF

Gumnuts, probably Tristania, from Bugaldie, New South Wales. (Specimen AMF. Magn.X 6.3)

PART TWO

PREHISTORIC ENVIRONMENTS OF AUSTRALIA AND NEW ZEALAND

Australia and New Zealand . . .
   The Antipodes . . . on the opposite side of the globe to Britain and Europe — the name implying the position of two feet, one in that familiar land and the other "a giant's stride" away in a strange and unfamiliar place, Australia (with New Zealand thrown in for good measure).
   Two islands . . . one an island-continent; the other almost a continent of islands, more isolated from other lands than any other country of comparable size.

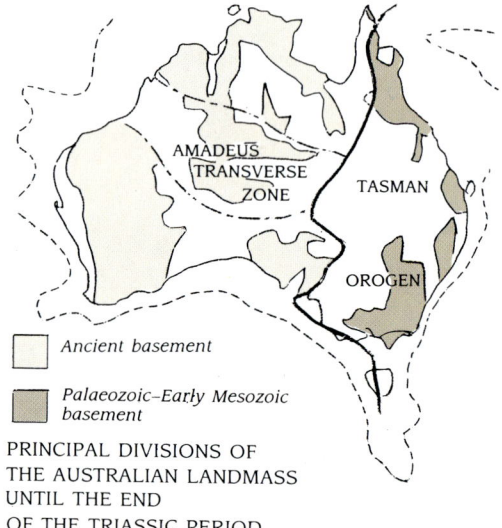

Ancient basement

Palaeozoic-Early Mesozoic basement

PRINCIPAL DIVISIONS OF
THE AUSTRALIAN LANDMASS
UNTIL THE END
OF THE TRIASSIC PERIOD

The mobile Amadeus Transverse Zone separated a northern and a southern block of stabilised land. The Tasman Orogen was a "growing region" which became progressively stabilised from west to east.

In terms of age Australia is a land of contradictions. Some of its rocks are among the oldest in the world, formed more than 4000 million years ago, and some of its ancient landforms have changed little since distant prehistoric times, yet in terms of geological time Australia, and Antarctica, are the youngest of the modern continents, having been the last to separate only about 45 million years ago.

Australia was part of the vast supercontinent of Gondwana for much of its history, and while connected to the other lands its position on the globe changed. In the course of time it moved from north of the Equator to South Polar regions. It was subjected to endlessly changing conditions throughout the 570 million years of the Phanerozoic Eon, and supported an ever-changing assemblage of plants and animals which evolved to suit the changing times. It long shared its fauna and flora with the other lands of the supercontinent, and only the Fossil Record of the last 45 million years is specifically Australian. The unique modern flora and fauna of Australia are the end-products of the evolutionary processes which started in the earliest prehistoric times and continued throughout the period of Australia-in-Gondwana. From the quota of Gondwanan plants and animals which were established in Australia at the time of its separation, there has been evolution-in-isolation. The biota has adapted to the specific environmental changes which have converted Australia from the rainforested land it was when it lay in polar regions, attached to Antarctica, to the driest vegetated continent on Earth.

The geological history of Australia is documented in textbooks and scientific papers, and is largely beyond the scope of this book. However, to create a framework for each geological Period an outline is included with a little detail about some of the major events and changes which affected environments. A series of Palaeogeographic Maps shows how the Australian landmass developed through time, and global reconstructions show how the world has changed through the ages.

The ancient land of Australia-in-Gondwana, 570 million years ago at the start of the Age Of Visible Life when our story begins, consisted basically of stable southern and northern blocks separated by the "Amadeus Transverse Zone" which was a mobile belt, not yet stabilised. The eastern half of today's continent was to be gradually added over succeeding geological Periods. This "growing region", known as the Tasman Orogen, had only become a stable, cohesive entity by early in the Mesozoic Era (about 230 million years ago). During the evolution of the continent, more stabilised land was added to the advancing eastern margin. Volcanic arcs operating to the east of the developing land migrated outwards as sedimentation and volcanism increased the size of the landmass. (Volcanic arcs form on the zones of interaction between adjacent plates.) Several plates were involved in the complex initial interactions on Australia's eastern margin at the start of Phanerozoic time. When the whole area now occupied by Australia had been stabilised, continuing activity created the outer edge of Gondwana which was to become the Lord Howe Rise and the Ancestral New Zealand landmass.

New Zealand has, in contrast to Australia, a comparatively short geological history as land. Its foundations were laid down under the sea in basins and troughs on the outer edge of Gondwana. While some small bits of land may have existed from time to time in the area early in the Mesozoic, the first substantial Ancestral New Zealand landmass was created by a series of tectonic movements which took place during the Jurassic Period, 208 to 144 million years ago. This ancestral landmass was much more extensive than modern New Zealand. The use of the term "Ancestral New Zealand landmass" is descriptive of a land area which existed briefly and is now mainly submerged. It extended from the Campbell Plateau off Antarctica in the south to New Caledonia in the north, and from the Lord Howe Rise in the west to the Chatham Islands in the east. After it severed its Gondwanan connection with Marie Byrd Land it was rapidly eroded and submerged, and today New Zealand, New Caledonia and Lord Howe Island are the major remnants.

In geological terms, New Zealand has not existed in anything approaching its present configuration until very recent times. Although usually considered as comprising only North Island and South Island, Stewart Island is also a part of New Zealand. Sir Charles Fleming, the famous New Zealand geoscientist, described New Zealand as "a changing archipelago" because of the scatter of small islands which are part of its history and present-day realm.

Modern New Zealand's unique, highly endemic flora and fauna are evidence of the isolation of its Gondwanan component of plants and animals. The small numbers of native species of plants and animals are the result of impoverishment over time. Many of the families and genera which it received from Gondwana did not survive the rigours to which they were subjected during the turbulent history of their land.

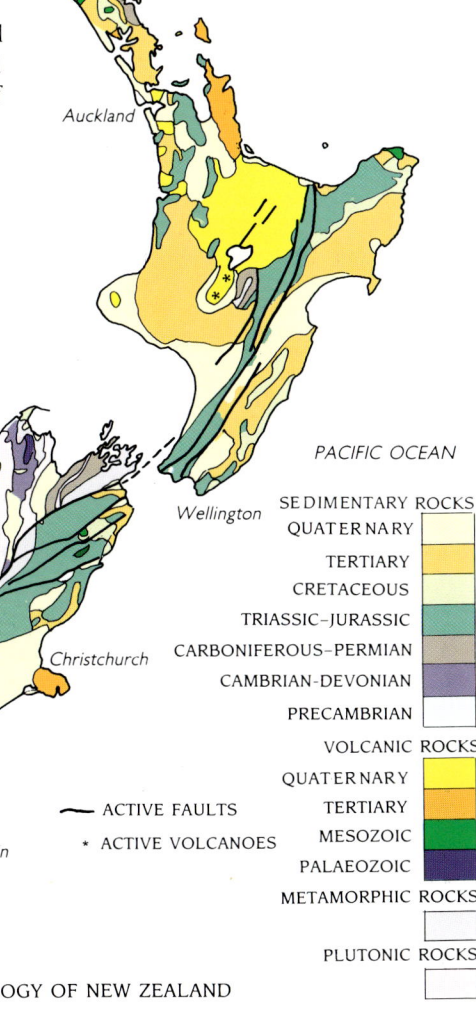

GEOLOGY OF NEW ZEALAND

SEDIMENTARY ROCKS
QUATERNARY
TERTIARY
CRETACEOUS
TRIASSIC–JURASSIC
CARBONIFEROUS–PERMIAN
CAMBRIAN–DEVONIAN
PRECAMBRIAN

VOLCANIC ROCKS
QUATERNARY
TERTIARY
MESOZOIC
PALAEOZOIC

METAMORPHIC ROCKS

PLUTONIC ROCKS

— ACTIVE FAULTS
* ACTIVE VOLCANOES

PACIFIC OCEAN
TASMAN SEA
ALPINE FAULT

Auckland
Wellington
Christchurch
Dunedin

THE PALAEOZOIC (ANCIENT) ERA

Gondwana was already a supercontinent with all its parts on one plate at the start of the Palaeozoic Era. It straddled the Equator but would soon change its position on the globe to become the Great Southern Land when the Tethys Sea opened and swung it southwards during the Carboniferous Period. Asia was represented by more than one plate during the Era — China, Siberia and parts of South-east Asia formed separate blocks which gradually moved towards each other and towards Euramerica and Gondwana resulting in the formation of Pangaea, the "one Earth", late in the Permian. (Pangaea was established as a stable landmass by early in the Mesozoic Era.)

In Late Palaeozoic times North America became welded onto the western margin of Europe approximately along the line where rifting would later occur to form the North Atlantic Ocean. At this stage Europe extended eastwards only as far as the region which is now occupied by the Ural Mountains. Sediments were being laid down along this eastern margin and would be raised up as the Urals when collision with the Asian block occurred in the Permian Period during the assembling of Pangaea.

Throughout the Palaeozoic Era the continental interior of Gondwana underwent considerable differential subsidence and uplift. Major basins were created and accumulated thick sequences of sedimentary rock.

The start of the Palaeozoic Era is delineated by the appearance of the easily visible Fossil Record, as a result of the evolution of shells and exoskeletons in marine Invertebrates. For the first 150 million years of the Era — comprising the Cambrian, Ordovician and most of the Silurian Periods — life was confined to the water. Evolution was rapid and most of the phyla of the Animal Kingdom were established during this time. The Plant Kingdom was represented by the Algae, and by the Late Silurian they were diverse and highly evolved, comprising all classes from microscopic, unicellular plankton to complex seaweeds of all the kinds (green, brown, red and coralline) which exist today. Land-plants evolved from an algal ancestor late in Silurian times.

The land world became an environment suitable for life when the concentration of oxygen in the air was sufficient to have created an ozone layer adequate to filter out enough harmful radiation to allow organisms to emerge from under the protection of water. Plant life in the sea had been responsible for establishing this situation, and now plants on the land would modify the harsh environment of a dead land world of rocks and sand to create environments with microclimates suitable for animals, where they could find food and shelter. Evolution took a mighty leap ahead when the environmental barrier between the realms of water and land had been breached.

Primitive land-plants established a vegetated zone on water margins, which enabled some animals to leave the water and start the invasion of the land world. Among Invertebrate phyla only the Arthropods were particularly suited to life on the land, and members of nearly all classes of Arthropods made the transition to terrestrial living.

The evolutionary adaptation of early land-plants to life on the land was a profoundly significant step, arguably equal to the evolution of green life in the beginning when the destiny of life on this planet was determined. Had plants not succeeded on the land, it is conceivable that their evolution might have proceeded no further than Algae, and Vertebrate evolution might have stopped with Fish. Both Algae and Fish were highly evolved and adapted to their aquatic lifestyles, and capable of occupying all the niches which a water environment offered.

Having aquatic ancestors suited and fine-tuned to living and breeding in water posed problems for land invaders. Indeed the story of evolution of the Plant Kingdom and of Vertebrates in the Animal Kingdom has been largely one of dealing with, and overcoming, the problems inherent in having such an inheritance. Not only body structure but also reproductive mechanisms had to be adapted to suit organisms to life on the land.

The environment in the seas is relatively constant and predictable in comparison with that on the land. Living under air instead of water, organisms had to face exposure to the Sun's heat and radiation, and to greatly increased diurnal and seasonal fluctuations in temperature, as well as exposure to wind and rain, flood and drought. Drying out of tissues became a serious issue. Fundamental changes had to occur before life on the land could succeed. These changes involved adaptations for breathing air instead of absorbing oxygen from water, the development of supporting structures for bodies no longer buoyed up by water, altering the methods of locomotion in animals, and developing new distribution mechanisms for spores and seeds in plants.

The changes required in reproduction were also significant. Algae have reproductive cycles based on free-swimming gametes and alternation of generations, characteristics which were a disadvantage on the land. The aquatic ancestral Vertebrates, Fish, shed their eggs and sperm into the water where fertilisation and development of offspring occurred independently of the parents — an arrangement obviously not viable on the land. Invertebrates making the transition to the land faced the same problems. The Arthropods were able to adapt because their structure was suitable and little change was needed, but other phyla could not establish terrestrial branches successfully.

Evolutionary adaptation to overcome the disadvantages of having aquatic ancestors for land-living organisms was along similar lines in the Plant Kingdom and in Vertebrates. For Vertebrates the evolutionary series was: from Fish with gills to the development of lungs for breathing air in the connecting-link Lungfish; to Amphibians suited for locomotion and living on land but having to return to the water to breed because of water-dependent larval stages; to Reptiles which perfected the egg and eliminated the need for free water; and to Mammals which retained the egg inside the body of the mother, giving protection and a better chance of survival to offspring and also offering the advantages of warm blood and of milk to nurture the young.

The Plant Kingdom solved the problems mainly with the development of vascular plants which had conducting and supporting tissues as well as specialisation of the plant body. Special organs were developed: roots for absorption and anchoring; stems with vascular tissues for the transport of water and prepared foods, and for strength to display other organs; leaves for photosynthesis; and sporophylls, cones, flowers and fruit to bear reproductive structures. Changes in reproduction were from plants which produced spores and had an alternation of generations (which required free water for fertilisation) to seed-producing plants which eliminated the alternation of generations and the need for free water. The spore-producing vascular plants — Horsetails, Ferns, Clubmosses — equate to the Amphibia; gymnospermous Seed Plants equate to the Reptiles, with the seed and the egg having broken the ties with the water; and the Flowering Plants (Angiosperms), with their seeds enclosed in a vessel, equate to the Mammals. However, the Flowering Plants and the Mammals did not appear until the following Mesozoic Era. The plants and animals of the Palaeozoic were all of the ancient groups which had not yet reached this highest level of evolution.

Glossopteris leaves from the Permian at Wingen in New South Wales. The Glossopterids evolved on the fringes of the ice in Gondwana as the world emerged from the Carboniferous–Permian ice age and lived in swamps which were features of landscapes during the Permian. (Mag.x 1.5)

CHAPTER 1

LIVING SEA, DEAD LAND

150 MILLION YEARS OF LIFE IN THE WATER
THE CAMBRIAN, ORDOVICIAN AND EARLY TO MID SILURIAN PERIODS
APPROXIMATELY 570 TO 420 MILLION YEARS AGO

A Trilobite from the Beetle Creek Formation, Queensland. Trilobites were early Arthropods, and are the most characteristic fossils of Cambrian times. (Specimen AMF. Magn.X 6.4)

To visualise the world of this far distant time . . .

Add to moon landscapes a pattern of oceans, lakes and rivers. Imagine rain falling on the rocks and boulders, and on the grits and sands of an erosion-eaten world. Brown run-off rivulets join browner rivers which braid and twist their threads across drab floodplains and pale deltas. Great stains of brown bulge into the fringing seas and drop their sterile burden — the residue of mighty mountains ground to mud and dust by the ravages of time.

See the winds blowing dust-storm clouds across deserted wastelands and find no green to rest the eye, no life to break the loneliness. An unimaginable emptiness.

Then turn away from the glare of silica and frowning rock and face the sea, cool and blue and inviting. The salt-laden breeze is refreshing and the waves roll up the fresh-washed beaches. The sinuous tidelines drawn upon the sand display the shells of strangely familiar creatures along with stranded jellyfish, bits of pumice and seaweed curling in the sun. Their message is: this planet is not dead, but keeps its life beneath the waters to wait until the day comes for new frontiers to be crossed and for a new, dry world to be invaded.

THE CAMBRIAN PERIOD

FROM 570 TO 505 MILLION YEARS AGO
DURATION: 65 MILLION YEARS

The evolution of shells and hard exoskeletons suddenly made animal life visible in the Fossil Record. After about 4000 million years of preparation our oxygenated world, with its bounty of living things, had been born. From this remote beginning in the dim past, evolution of the Plant and Animal Kingdoms was under way and would culminate in the communities of plants and animals of the present day. For we, and all life, came from the sea and we carry its salts in our blood some half a billion years later.

In the relative safety of the seas, protected from the elements which ravaged the dead, eroding and hostile land, the ancestral members of the different groups (phyla) which now inhabit Earth came into being. Marine communities established within Cambrian times were not unlike those that we know today. Some species were mobile and scavenged the sea bottom, some were sedentary and obtained their food by filtering water, some burrowed and ingested mud, some secreted lime and made reefs, and some were free-swimming.

Trilobites, among the earliest Arthropods, are the characteristic and most interesting animals of the Period. The basis of the food chain then, as now, was plants, though at this time the Plant Kingdom was represented by members of only one phylum, the Algae. These Algae were of different sorts, from the microscopic single-celled plankton to seaweeds and coralline-encrusting varieties.

An assortment of hard, shelly parts of animals remained when a Mid Cambrian limestone specimen from Woolomin in the Tamworth district of New South Wales was dissolved in a bath of weak acid. Shells of Brachiopods and Molluscs and fragments of Trilobites in this 540-million-year-old sample testify to the abundance of marine life. It was the appearance of hard parts like these that facilitated the separation of the Age Of Hidden Life (the Cryptozoic Eon) from the Age Of Visible Life (the Phanerozoic Eon) in the Geological Time Column. The base of the Cambrian Period was determined by the appearance of shells in the Fossil Record at 570 million years ago. (Specimen MQU. Magn.X 4.0)

Much of the world's land lay in low and middle latitudes during the Cambrian Period, implying that climates were hot worldwide. There is also evidence that some areas were affected by aridity. However, temperature gradients may have been considerable at times, as rocks in North Africa show evidence of glaciation, which in turn suggests that seas would then have been cooler. With life confined to the waters, this information is important. Moreover, when there is only fragmentary evidence of one cold interval in a Period as long as the Cambrian, it is always possible that there were other cold episodes whose traces have been obliterated by erosion.

AUSTRALIA IN THE CAMBRIAN PERIOD

During Cambrian times Australia lay in the Northern Hemisphere. Part of the present-day coastline of Western Australia was on the Equator and the rest of the land lay between 0° and 30° North. Globally, sea levels were high. Late in the Period part of the continent was inundated when an embayment into the Centre increased the areas of shallow marine environments. Invertebrate marine life was abundant, with bottom-dwelling (benthic) organisms predominant.

Much of Australia was low-lying, with hilly ground confined to the southern part of the Northern Territory and to the Kimberleys region of Western Australia. Under locally arid conditions, shallow embayments of the epicontinental seas were cut off from the open water. The waters of these bays evaporated and phosphorites, gypsum and other evaporite deposits formed.

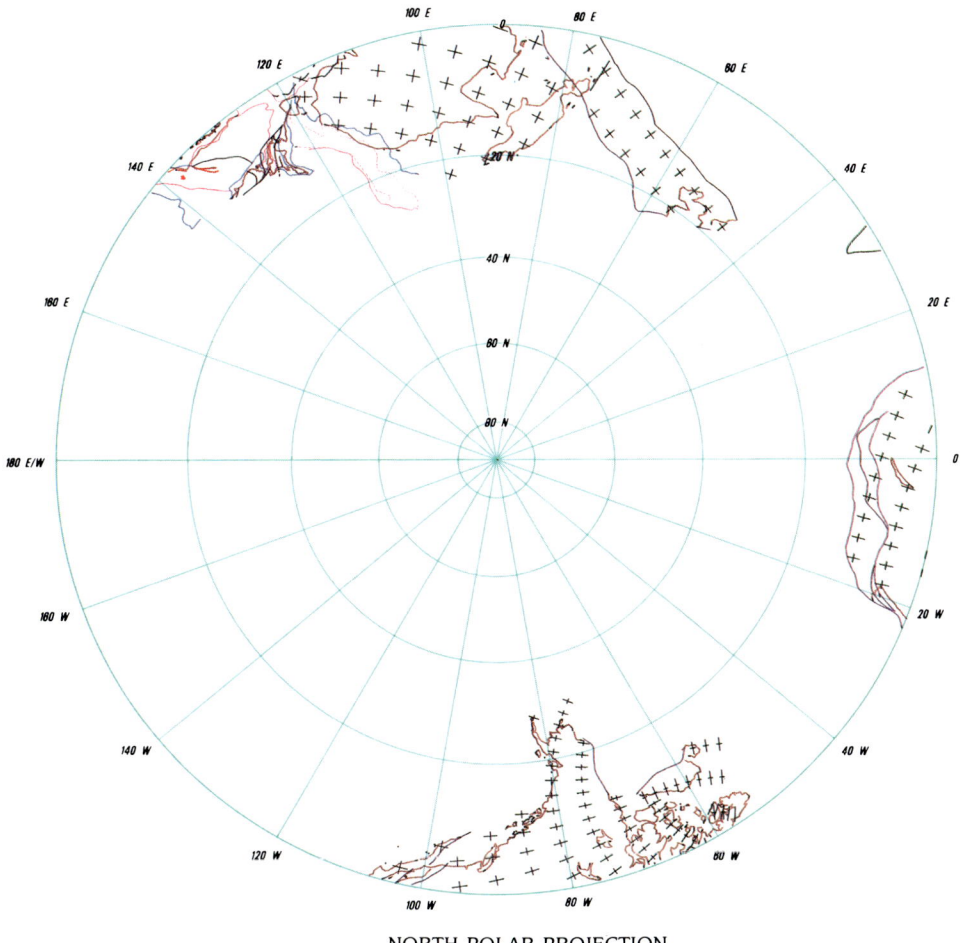

NORTH POLAR PROJECTION

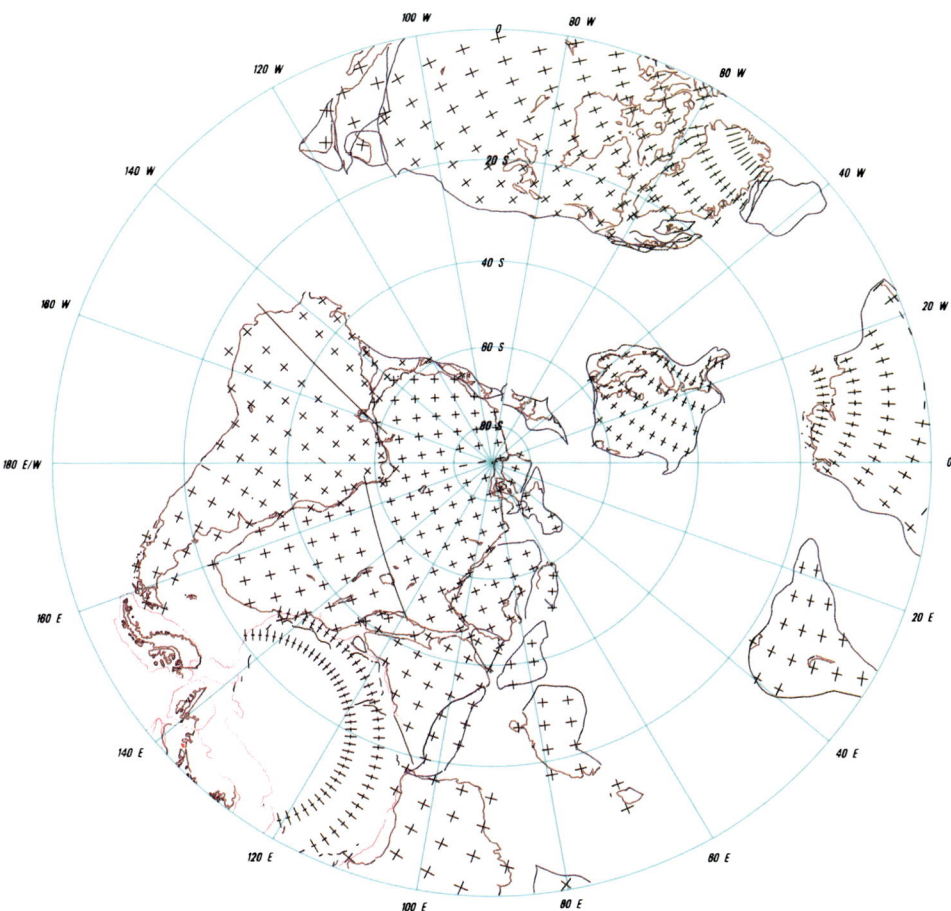

SOUTH POLAR PROJECTION

An outcrop of Early Cambrian Beetle Creek Formation rocks in the Mt Isa district of Queensland. Beetle Creek is so named because of the abundant Trilobites (''Beetles'') which occur in the rocks there.

POSITION OF LANDMASSES IN RELATION TO THE NORTH AND SOUTH POLES IN THE CAMBRIAN PERIOD, 520 million years ago

A Trilobite, Redlichia forresti, *from the Early Cambrian Linnekar Limestone in the Ord Basin, Western Australia. (Specimen WAM. Magn.X 4.7)*

Rusophycus, a Trilobite trace fossil produced by the animals feeding in the mud on the sea bottom. Their legs made the scratch marks on the mounds of mud, which they pushed up as they turned the silt over to sort through it for food particles. (Specimen MMF. Magn.X 0.8)

Extremely small Molluscs with shells only a few millimetres in width occur early in the Cambrian Period. The evolutionary trend was towards an increase in size over time. The tiny shells seen here enclosed in 0.5 cm × 0.5 cm segments were dissolved out of the Mid Cambrian Currant Bush Limestone from the Georgina Basin in the Northern Territory. (Specimen MMF. Magn.X 1.5)

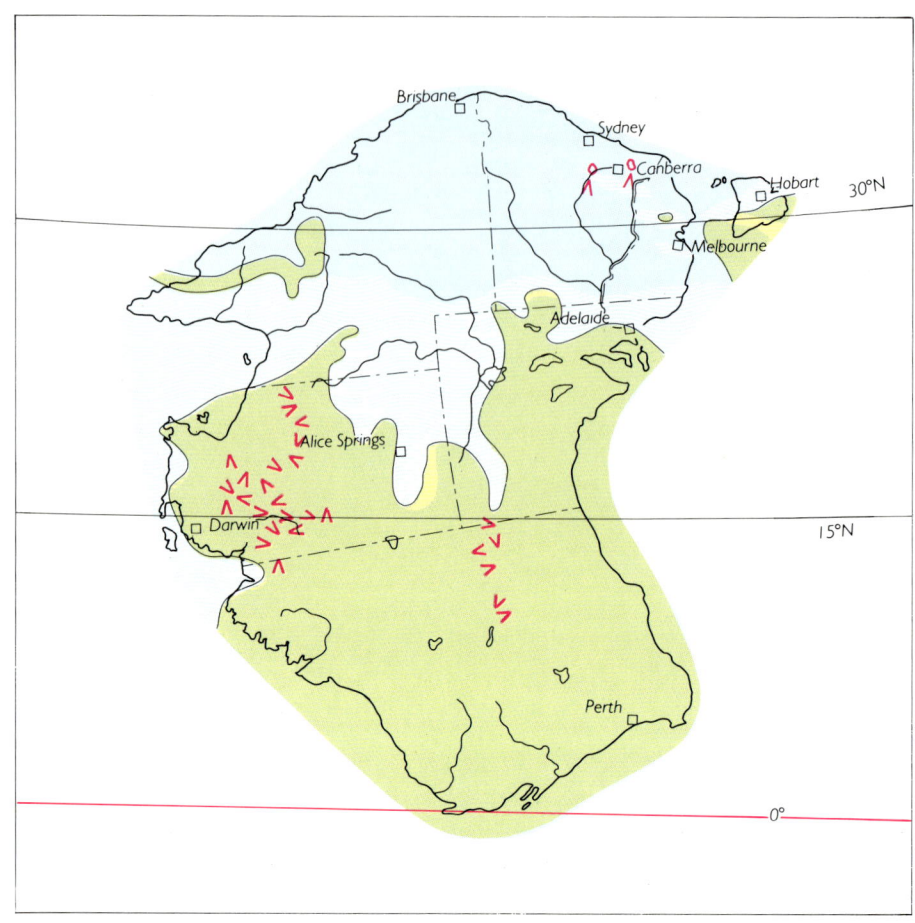

PALAEOGEOGRAPHY OF THE LATE CAMBRIAN
520 million years ago

Massive volcanic activity occurred in the northern part of the Northern Territory and in the Great Victoria Desert region of Western Australia. Basaltic flows formed enormous sheets covering great areas. The Deccan Traps in India today are basaltic flows on a similar scale to the Cambrian flows in Australia. Volcanic activity on a much smaller scale also occurred in New South Wales, Victoria and Tasmania.

Towards the end of the Cambrian, seas over the Gulf of Carpentaria region and those in the Centre retreated due to uplift and folding in a zone across the continent northwards through Adelaide. This tectonic event is known as the Delamerian Orogeny.

The Late Cambrian and Ordovician times are of great significance in the story of petroleum in Australia. Reservoir rocks and probably source rocks of the Mereenie Oilfield (the largest on the mainland) and the Palm Valley natural gas field were laid down in the Amadeus Basin. Smaller, non-commercial pools of condensate and gas are also located in the area. As well, exploration companies hope that the Canning Basin in Western Australia and the Arafura Basin off northern Australia may prove to have similar commercial potential.

The Plant Kingdom in Cambrian times was represented by Algae. As Algae have no readily preserved tissues, their fossil record is very poor other than for the lime-secreting types which were involved in reef formation.

In the Australian seas reefs were formed by Archaeocyathids, Coral-like organisms whose true affinities were long in doubt. Recent research, however, indicates that they are early Sponges and that their biology had

THE BURGESS SHALE FAUNA

The Burgess Shale is a world-famous Mid Cambrian rock unit in western Canada where an assemblage of delicate and soft-bodied animals is so well preserved that all the minute detail of their structure is visible. It is a rare enough occurrence to have delicate animals preserved as fossils at all, and here in the Burgess Shale the amazingly weird creatures are a particular source of wonderment. The fauna is so different from anything one would have expected of Cambrian animals, if one's knowledge had been gained from a study of the "normal" fossils of shells and hard exoskeletons. The limitations of the Fossil Record are clearly revealed by the discovery of this Burgess fauna. Was there a whole world of other sorts of animals which have left no record throughout geological time?

The preservation of the Burgess Shale animals was the result of the unique conditions which prevailed when they were entombed in fine sediment about 530 million years ago. An area enclosed by reef became filled to overflowing with fine silt. When it could hold no more, slumping occurred, and an assortment of bottom-dwelling animals which lived below the reef were rapidly buried. Any creatures swimming about where the slumping occurred were also enveloped.

Typical Cambrian forms — the Trilobites, Brachiopods and other shelly creatures which characterise other Cambrian localities — make up about 20 per cent of the Invertebrate genera in the fauna. The rest of the 120 species are remarkably well-preserved, soft-bodied creatures representing eight known and 10 or more unknown phyla. The highly specialised adaptations shown by these animals are in marked contrast to the lack of specialisation which characterised the Ediacaran soft-bodied fauna of 70 million years before.

Marella splendens, a delicate creature with long, horn-like appendages on its head — an example of the strange and wonderful fauna found in the Burgess Shale. (Specimen AMF. Magn.X 3.6)

The "atypical" 80 per cent of the Cambrian fauna seen in the Burgess Shale includes a Peripatus-like animal (then aquatic, while its modern relatives are terrestrial "living fossils") and 30 other species of Arthropods. There are seven species of unsegmented marine Worms, six Annelid Worms, and a Chordate (Vertebrate ancestor) called Pikaia. They are all representatives of groups of animals which are alive today. As well, an assortment of other creatures are exclusive to the Burgess Shale, and some of them are quite bizarre. One genus has been named Hallucigenia. It walked on seven pairs of stilt-like legs and had seven tentacles on its back, each possibly ending in a mouth.

The Burgess Shale fauna raises the question: were its animals widely distributed at the time and simply not preserved except under the most unusual circumstances? Isolated examples of Burgess-like animals have been found elsewhere in the world, and it seems likely that similar habitats supported similar faunas of delicate animals.

Peripatus, a "living fossil" whose ancestors were present in the Burgess Shale fauna of 530 million years ago.

DENSEY CLYNE

previously been misunderstood. An earlier assumption that Archaeocyathids became extinct at the end of the Cambrian Period is thus incorrect. Their evolution into easily recognised modern-style Sponges accounts for their disappearance from the Fossil Record.

Trilobites (Arthropods with exoskeletons composed of chitin), mainly of the bottom-dwelling and burrowing type, and Brachiopods (Lamp-shells) dominated a diverse Invertebrate fauna. Evolution of creatures to fill all the niches proceeded swiftly, and free-swimming, sedentary and burrowing members of all the phyla alive today were soon evident. Gastropod and Bivalve Molluscs were present throughout the Period. Some of the early ones are very small, with shells only a few millimetres in width, but during the evolution of the phylum there has been a tendency to an increase in overall size. By late in the Period the first large predators, Cephalopods, appeared. These animals are relatives of Squid, Octopus and *Nautilus*, and are also members of the phylum Mollusca.

Other Cambrian faunas include Echinoderms and Conodonts — minute phosphatic, tooth-like organs — which range from Cambrian to Triassic times. Conodonts are useful for correlating fossil horizons because they evolved rapidly and are often abundant. Recently, some Conodonts have been found inside the remains of soft-bodied creatures of unknown affinity — Conodontophorids — and are known to have been part of their feeding apparatus. They were called "problematica" of unknown origin before then.

The first Protistans to secrete hard tests, Foraminifera and Radiolaria also appear during the Cambrian.

It is obvious from the degree of complexity exhibited by Cambrian animals that much evolution had occurred during the Cryptozoic Eon, but the production of hard parts made the start of Cambrian times look like an explosive time for evolution. The evolution of so many different types of Invertebrates was a response to the availability of a wide range of ecological niches. Many of the animals were pioneering experiments by various groups and were later to be supplanted by better-adapted and more efficient organisms. A few groups appear to have been favoured at the expense of others which became extinct. Late in the Period, Ostracods and Bryozoans make their appearance in the Fossil Record.

NEW ZEALAND IN THE CAMBRIAN PERIOD

Cambrian deposits in the Cobb Valley, in the north-western Nelson region of South Island, are the oldest accurately dated geological formations in New Zealand. Some probably associated sequences occur in small areas in the south-western Fiordland. From the evidence of their thickness and texture, all the deposits appear to have accumulated in a shallow marine basin not far from land. The Tasman Formation of the Cobb Valley has a fauna of Trilobites, Brachiopods, Sponges and Ostracods which shows affinities with the Cambrian faunas of eastern Australia.

A great deal of volcanic activity occurred in Cambrian times, with great outpourings of basaltic and andesitic lavas. As well, some rocks of Cambrian age were formed from the ashes and rock fragments ejected by volcanoes. The volcanic activity appears to have been concentrated in the Early and the Late Cambrian, with much less in the middle of the Period.

Early to Mid Cambrian Monoplacophoran shells. Until recently these primitive, ancestral-type Molluscs were known only from the Cambrian, and were therefore thought to have become extinct 500 million years ago. Deep-sea dredging by the Danish "Galathea" expedition brought to light a "living fossil" representative of the group which has been named Neopilina. (Specimen MQU. Magn.X 4.0)

Early Cambrian Archaeocyathid limestone from the Mt Scott region of South Australia. The organisms are seen here in cross-section, with their septate wall structure and hollow centres in evidence. (Specimen AMF. Magn.X 3.1)

An Archaeocyathid Limestone from South Australia. Archaeocyathids are found only in the Cambrian Period. They were ancestral Sponges which were so abundant that their solid calcium carbonate skeletons formed reef limestones. At the end of the Cambrian they were replaced by primitive Sponges of more "normal" form and structure. (Specimen AMF. Magn.X 4.9)

Archaeocyathids seen in longitudinal section, showing the elongated, flask-like shape of the organisms. (Specimen AMF. Magn.X 4.3)

Large numbers of very small Agnostid Trilobites are found in the Beetle Creek Formation of Queensland. (Specimens AMF. Magn.X 3.7)

73

Ordovician Brachiopods (Lamp Shells) from Nelungaloo, New South Wales. Brachiopods have two calcareous shells and resemble Bivalve Molluscs, but they have differences in anatomy and symmetry which require their classification in a separate phylum. Brachiopods range from the Cambrian to the present day and were particularly abundant in the Palaeozoic Era. (Specimen MMF. Magn.X 2.6)

THE ORDOVICIAN PERIOD

FROM 505 TO 436 MILLION YEARS AGO
DURATION: 69 MILLION YEARS

Life in the Ordovician Period was still confined to the water and there was a great increase in the diversity of living things. Early Fish, the first Vertebrates, appeared in the middle of the Period. High sea levels worldwide caused epicontinental seas to encroach on the land and the shallow-water environments which resulted supported rich assemblages of animals. The seas were warm and Corals were abundant. Late in the Period there was a global cooling event, leading to an ice age which persisted into the start of Silurian times. Sea levels fell as the ice volume increased.

Australia was situated in equatorial regions and was little affected by the ice age. Its waters remained warm and there was no rapid change in conditions for the organisms that lived in them.

Graptolites (free-swimming, colonial organisms) are the characteristic Invertebrates of this Period.

Landmasses of the world in Ordovician times comprised the supercontinent Gondwana, of which Australia was a part, and a number of comparatively small blocks of land which were evolving into the continental masses of the Northern Hemisphere. The positions of land on the globe changed, with Gondwana swinging away from the Equator so that by late in the Period its northern South American margin lay further into polar regions. Australia, however, still lay north of the Equator in much the same situation as in Cambrian times; the movement of the supercontinent was greatest at its other extremity.

The climate seems to have been hot in the Early and most of the Late Ordovician, cooler in the Mid Ordovician and coolest towards the very end of the Period when there was polar ice which persisted into earliest Silurian times. Lands like Australia, lying in low latitudes, were more affected by the sea level changes which resulted from ice formation and melting than they were by temperature variations. Sea levels were at their lowest at about 440 million years ago, indicating the peak of the ice age.

The Period had a rich shallow-marine fauna comprising members of the same groups which were abundant during the Late Cambrian. In addition, ancestral Horseshoe Crabs and Eurypterids make their appearance in the Record. Crinoids are also found from the Early Ordovician onwards.

AUSTRALIA IN THE ORDOVICIAN PERIOD

Throughout the Ordovician Period, Australia lay between the Equator and 30° North. The Fossil Record shows that the first Vertebrates, Fish, appeared by the middle of the Period. Sea levels were high, as in the Cambrian Period, falling only during ice cap formation at the end of the Period. The seas remained warm and supported a diverse fauna.

The Delamerian Orogeny, the mountain-building episode which had begun in the Late Cambrian, continued to raise high ground in a zone which extended from western Tasmania in a north-north-westerly direction through western Victoria to north-central South Australia. Erosion of this high ground was rapid, implying high rainfall.

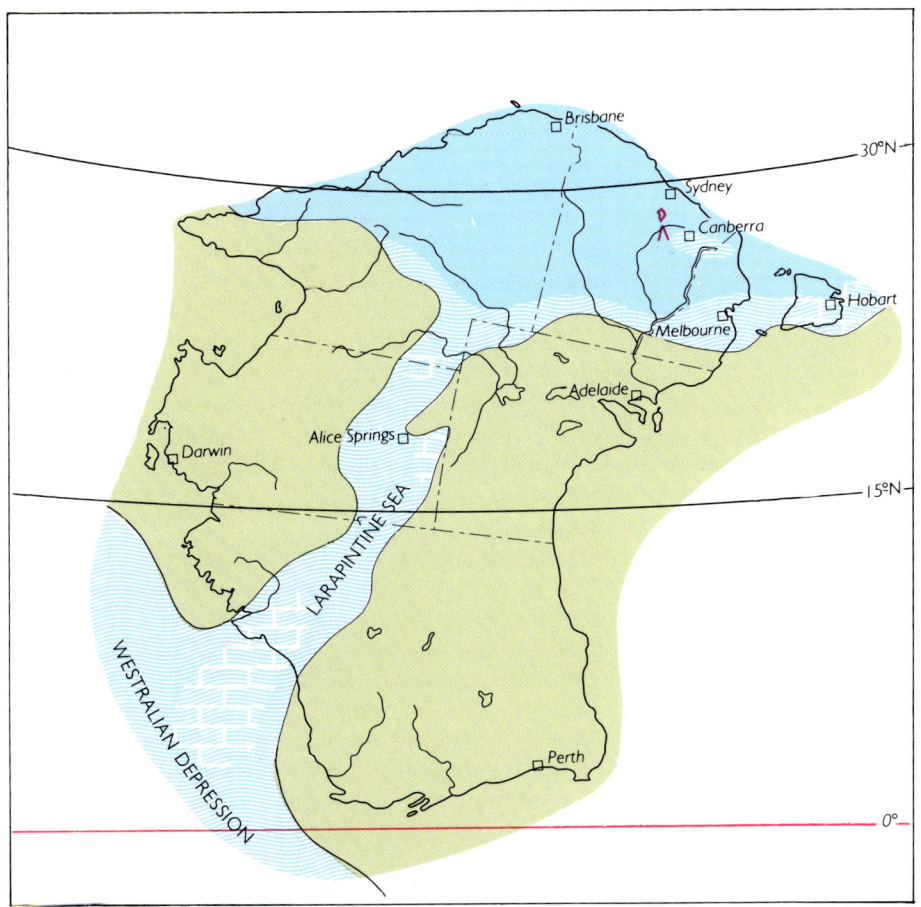

PALAEOGEOGRAPHY OF THE MID ORDOVICIAN
460 million years ago

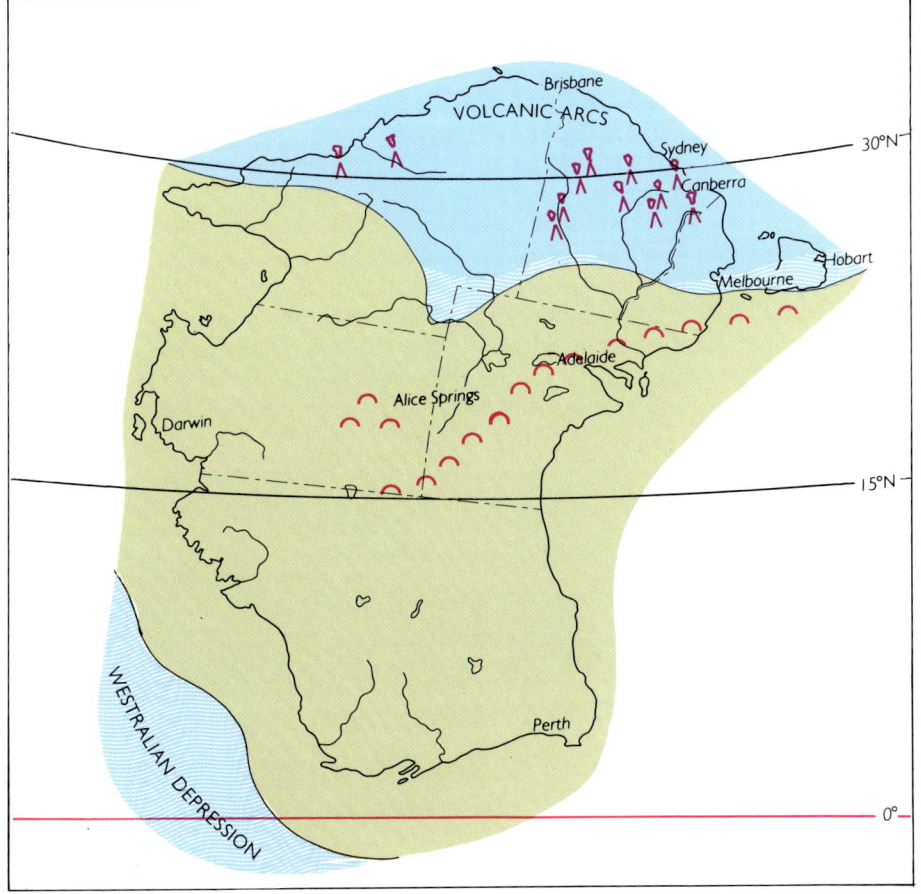

PALAEOGEOGRAPHY OF THE LATE ORDOVICIAN
445 million years ago

CONODONTS

Conodonts are minute, phosphatic, tooth-like organs which range in size from microscopic to about 3 millimetres long and often occur in great abundance in rocks of Cambrian to Triassic age. They can be extracted from limestone intact because the phosphatic material they are made from can resist the weak acid which is used to dissolve the limestone.

Conodonts are useful in correlating zones and determining the age of rocks, because they evolved rapidly and certain assemblages are characteristic of certain time zones.

The nature of Conodonts has only recently been determined. For a long time they were simply considered "problematica" whose source and function were unknown. Recently a specimen of a Worm-like creature about 4 centimetres long and 2 millimetres wide has been discovered with Conodonts within its mouth area. Indeed, this creature, *Clydagnathus*, had more than one type of Conodont in its head, and the phosphatic tooth-plates seem to have been used as teeth or as part of a filter mechanism. Such soft-bodied, Worm-like creatures must have been extremely abundant through the ages to account for the vast numbers of Conodonts which occur as fossils.

Conodonts were among the first hard-parts to appear in the Cambrian Period.

JOHN PICKETT

Conodonts, microscopic phosphatic tooth plates. Their range is from the Cambrian to the Triassic.

The western sector of the Amadeus Transverse Zone, which had been elevated above sea level during Cambrian and earliest Ordovician times, began a long history of subsidence during the Early Ordovician, eventually resulting in the formation of the Canning Basin. Seas advanced into this Canning depression by the Mid Ordovician, and joined up with an embayment from the eastern margin of the continent, effectively bisecting the Australian landmass. This new epicontinental seaway is called the Larapintine Sea, and it in turn joined up with a narrow channel that extended along Australia's north-western margin — the first part of the Westralian Depression. The land on one side of the Larapintine Sea, comprising northern Australia, was flat and featureless. That on the other side, comprising the southern half of Western Australia together with South Australia, was more rugged.

With the first movements of the Alice Springs Orogeny late in the Period, the epicontinental seas started to regress. The Larapintine Sea gradually disappeared, and many of the deposits which had accumulated in it were uplifted and eroded. Thus the Amadeus Transverse Zone ceased to exist as an entity. Where it had previously opened eastwards to the Tasman Orogen, it now opened westwards to the Westralian Depression.

Deep-water sedimentation continued along the eastern continental margin. Volcanic activity became more widespread in what is now eastern Australia, building up a foundation of volcanic rocks in episodes which represented another mountain-building phase — the Lachlan Orogeny.

The wealth of shallow-water environments during Orodvician times promoted a rich and diverse fauna. There was a great increase in the number of suspension-feeding animals taking their food from the water. Brachiopods, Bryozoans, Molluscs and Echinoderms all underwent rapid evolution.

Graptolites characterise the Period, and their rapid rate of change makes them valuable tools in stratigraphy. There is a succession of different types through time and they can be used to subdivide the Period into zones, and to correlate geographically separated strata.

Trilobites were also abundant and diverse. One very large species, up to half a metre in length, has been found in the Stairway Sandstone which was laid down in the Larapintine Sea. Trilobite nests or burrows are often preserved as trace fossils known as *Cruziana* and the scratch-marked piles of mud which result from their feeding activities are known as *Rusophycus*. Predatory Cephalopods, including straight-shelled Nautiloids were abundant, and the presence of Corals indicates that water temperatures were warm.

A small Starfish (phylum **Echinodermata***) from South Australia. (Specimen U.NSW. Magn.X 4.1)*

An Ordovician Sponge, as yet undescribed, from Cheesemans Creek, New South Wales. (Specimen MMF. Magn.X 2.2)

The first armoured Fish, an Ostracoderm called *Arandaspis*, swam in the shallow and well aerated waters of the Larapintine Sea. It and three of its relatives were preserved in the Stairway Sandstone. The Vertebrates had arrived!

A most interesting and controversial group of organisms with very complex structure, the Receptaculitids, also first appear in the Ordovician Period and range through to Devonian times. They are now believed to be corraline Algae, having long been classified as "problematica".

Trilobite heads from Late Ordovician rocks, about 500 million years old, at New Durran, New South Wales, where fossils like this example are abundant. Incomplete Trilobite fossils are much more common than those of whole creatures, and head shields occur almost to the exclusion of other body parts at the New Durran locality. (Specimen MMF. Magn.X 2.0)

NEW ZEALAND IN THE ORDOVICIAN PERIOD

Well-defined Ordovician rocks, characterised by shelly and Graptolite faunas, are found in the Nelson region and in the south-western part of Fiordland in South Island. The shelly faunas are meagre, comprising only a few Corals and Trilobites. The Graptolite successions correspond to the Victorian Graptolite faunas of Australia.

There is evidence that the Ordovician sediments were deposited in an offshore basin, and the types of sediment found indicate that there was much less volcanic activity than in the previous Period. Fewer strata contain the ashes and debris of volcanic activity. The nature of the sediments also indicates that they were derived from areas of low relief, having characteristics consistent with low-energy environments.

Tectonic events appear to have been confined to the gentle sinking of a shallow basin, the axis of which was located to the west of the present South Island. (The folding of the Ordovician sediments probably occurred in later Palaeozoic times.)

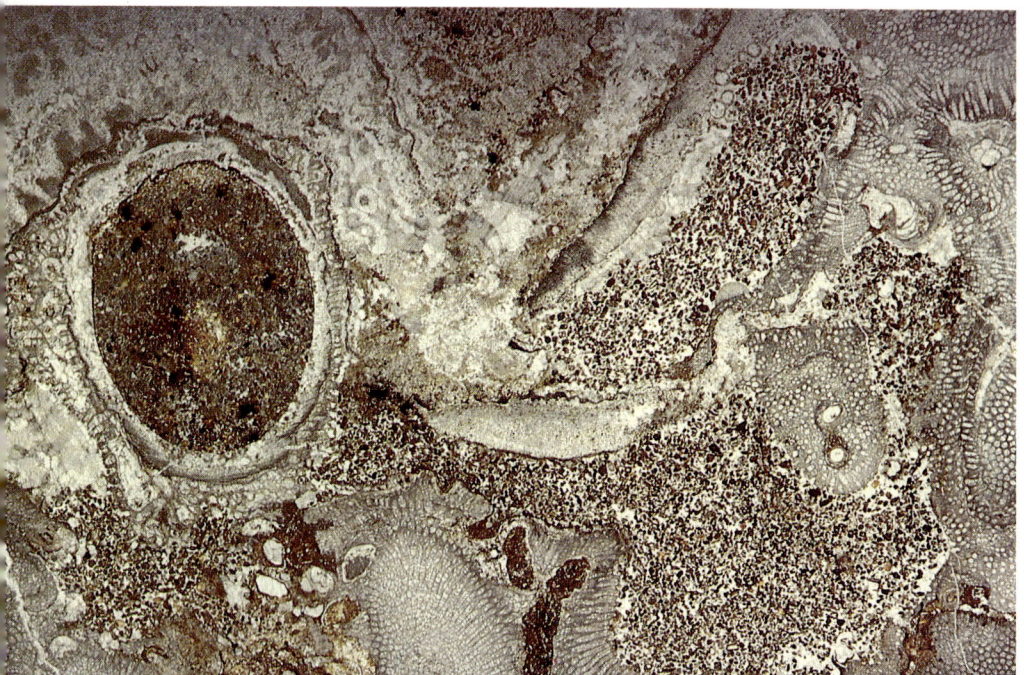

THIN SECTIONS OF ORDOVICIAN CORALS

A Stromatolite, Cliefdenia (round, in cross-section), and a Bryozoan (delicate lace-like pattern). These coralline organisms contribute to the formation of reef structures.

OPPOSITE PAGE:

A Tabulate Coral, Nyctophora sp.

A Rugose Coral, Plasmoporella inflata.

A cross-section of a Tabulate Coral, Nyctophora (centre, solid with pores), and a Solitary Coral, Hillophyllum (right). (Specimens MMF.)

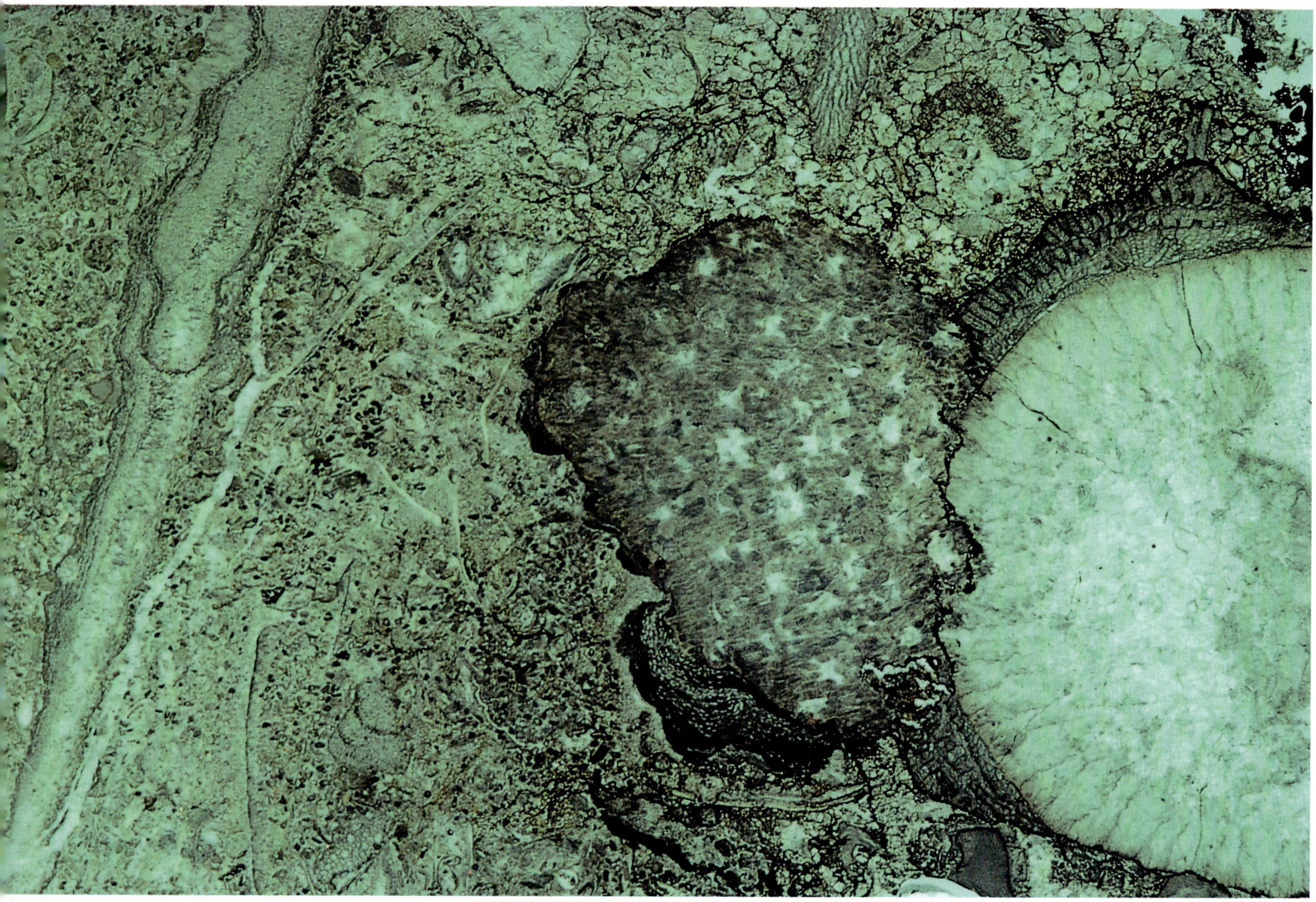

ORDOVICIAN GRAPTOLITES FROM
VICTORIA

*Graptolites from Dixons Quarry and Spring
Valley, Victoria.*
*Graptolites are an extinct class of colonial
marine organisms. Each consisted of a
number of chitinous cups which housed the
individual polyps, arranged along a stem.
The range of Graptolites is from the
Cambrian to the Carboniferous. (Specimens
MQU. Magn.X 1.5)*

THE EARLY AND MID SILURIAN PERIOD

FROM 436 TO 420 MILLION YEARS AGO
DURATION: 16 MILLION YEARS

During Early and Mid Silurian times life was still confined to the water. Coral reef formation was active in the warm seas, Trilobites were still abundant though starting to decline, and Brachiopods and Graptolites continued to abound.

On land, the ultraviolet solar radiation was being increasingly filtered out by the ozone layer. With the steady increase in oxygen in the atmosphere, the ozone layer was becoming denser and approaching a point where it would effectively reduce radiation to a level acceptable for living organisms unprotected by water

Land masses of the world comprised: Gondwana, with Australia straddling the Equator and the rest of the supercontinent in the Southern Hemisphere with the South Pole on North Africa; North America, Scandinavia and a bit of Europe united, and bisected by the Equator; and a number of separate blocks in the Northern Hemisphere which would later form Asia.

After the ice age on the Ordovician to Silurian time boundary, there was a global rise in sea level. In North Africa, where most of the glaciation had occurred, the rising sea levels caused the establishment of a shallow epicontinental sea. This sea spread over the Sahara, depositing organically rich sediments on top of the sediments which had been deposited by the ice sheets. Being highly porous, glacial sediments form ideal reservoir rocks, and the combination of marine source rocks and glacial reservoir rocks established the basis of the North African oilfields.

Small Trilobites from the Silurian at Bowning, New South Wales. (Specimen MMF.Magn.X 1.1)

AUSTRALIA IN THE EARLY AND MID SILURIAN

Australia still lay in near-equatorial latitudes during Silurian times. In the earliest part of the Period, while the ice age was at its most intense, the climate was probably cool. Sea levels were at their lowest at the start of the Period, and resumed their rising and high levels after the ice melted. Mid Silurian climates were warm to hot worldwide and there was no polar ice. Australian evidence supports the presence of warm waters on all its oceanic margins.

The Alice Springs Orogeny of central Australia, which had started in the Late Ordovician, proceeded. Mountain-building movements also occurred along the eastern seaboard, in the continuing Lachlan Orogeny. Limestones formed from reefs which had flourished in the shallow seas during marine incursions were among the strata uplifed and eroded during this tectonic episode. Some of the reefs were formed by Stromatoporoids, not true Corals. The famous caves at Bungonia and Jenolan in New South Wales and at Chillagoe in Queensland are remnants of these extensive limestone formations, in which subsequent weathering has hollowed out caves and dripping water has created the stalactites and stalagmites which make them natural architectural marvels.

A complex fracture system developed along the south-eastern margin of the continent, resulting in the subsiding of the Darling and Adavale basins. On the north-eastern margin a series of depressions and troughs developed.

Offshore from a fairly narrow shelf in the south-eastern sector of the continent, a complex pattern of troughs, shelves and elongated islands

Crinoid stems from the Late Silurian at Quedong, New South Wales. (Specimen MMF. Magn.X 1.4)

OSTRACODS

Ostracods are small Crustaceans sometimes known as "Seed Shrimps". They are laterally compressed and their bodies are enclosed in two-valved shells which are hinged along their dorsal surfaces (backs).

Ostracods have a long and well-documented history from Cambrian times onwards. They have adapted to different lifestyles and are found in ocean plankton, living on the seafloor, in freshwater ponds and even in damp soil. Early forms were benthic (bottom-dwelling).

Ostracods are second only to the Foraminifera in their abundance in the Microfaunal Fossil Record, having been major components of shallow-water marine faunas. They are particularly useful in the correlation of zones in marine strata, and as indicators of ancient shorelines and the relative depths of seafloor.

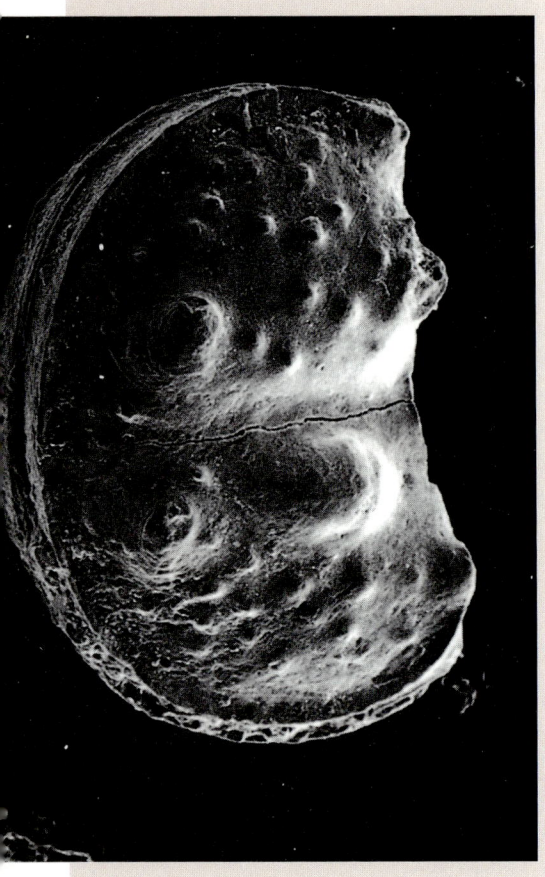

PETER JONES

PALAEOGEOGRAPHY OF THE LATE SILURIAN
420 million years ago

formed, their positions commonly changing with time. Volcanic rock was intruded in this area, sometimes forming islands in shallow water and sometimes cooling below the surface to form granite batholiths. Sedimentation continued in the deeper troughs, while shelf areas were sites for deposition of sandstones, shales and limestones. The limestones contain warm-water rugose and tabulate Corals, Trilobites and Molluscs.

Graptolites, which characterised the previous Ordovician Period, continued to be abundant during the Silurian. As well as being useful in dividing the Period into zones they also, because of their cosmopolitan aspect, enable correlation with European stages. Stromatoporoids (reef-forming, corraline organisms), Bryozoa, Crinoids, Starfish and abundant Ostracods were also present. The Eurypterids (Sea Scorpions) were at their most abundant and there were Horseshoe Crabs, Sponges and Annelids. The Annelids are not found as fossils because they are soft-bodied and lack hard tissues suited to fossilisation, but their presence is confirmed by trace fossils of their burrows and worm tubes, discovered notably at famous Western Australian localities.

Eurypterids were the first creatures to leave their footprints on land. Their trackways across rock platforms found in Western Australia show the considerable size which these creatures attained. They had probably been scavengers on the tidelines long before plants moved onto the land and made a living zone between the water and the dry, inhospitable world beyond.

NEW ZEALAND IN THE EARLY AND MID SILURIAN

No rocks of Silurian age have been recognised in New Zealand. If any still exist, they were formed from sediments accumulating in an offshore basin on the outer edge of Gondwana.

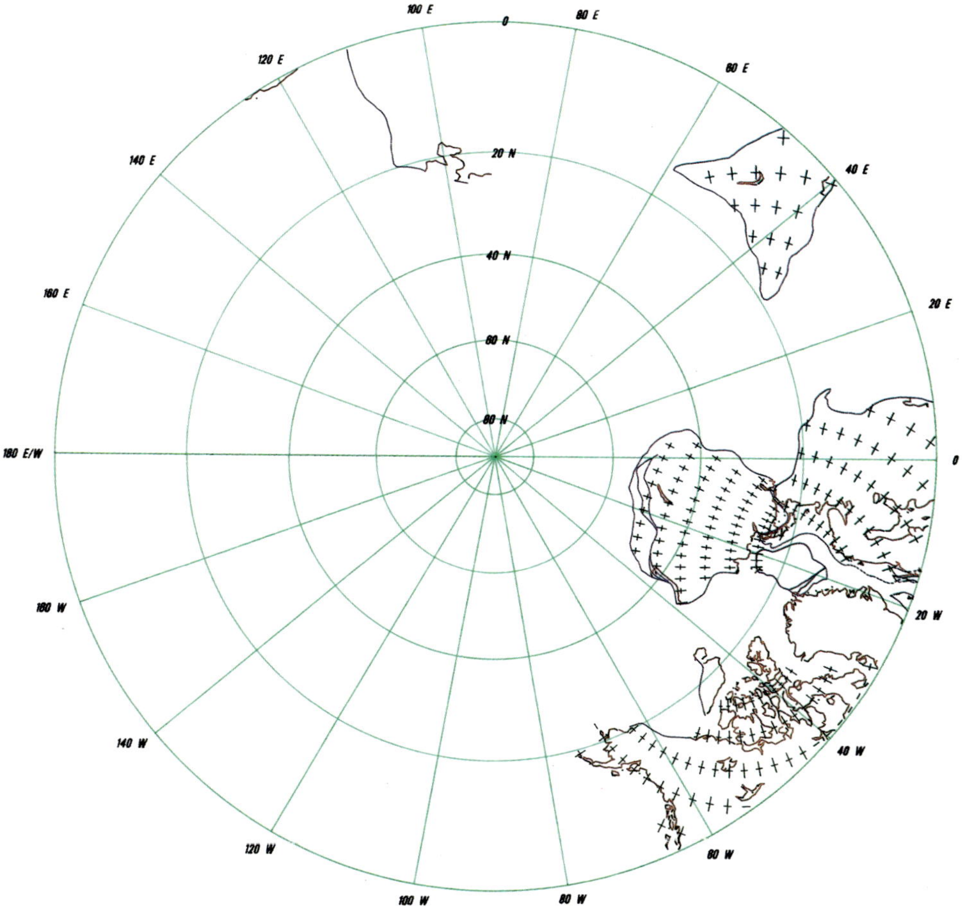

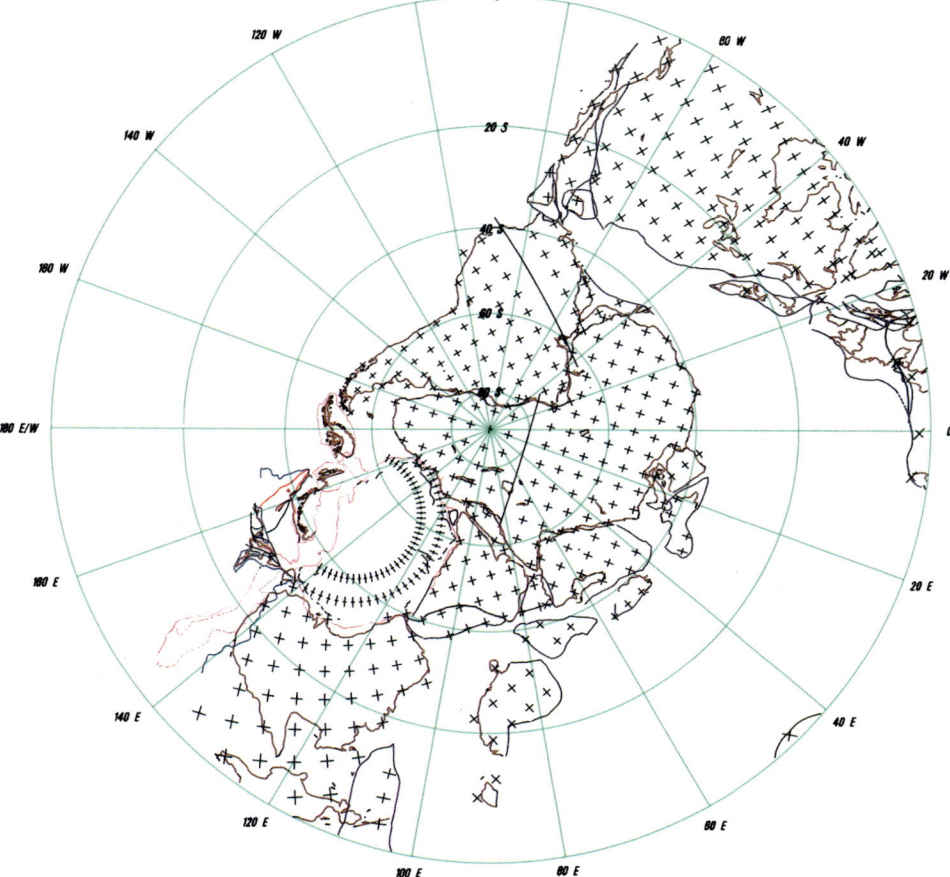

NORTH POLAR PROJECTION

SOUTH POLAR PROJECTION

Acanthohalysites australis, a Chain Coral from Forbes, New South Wales. (Specimens MMF.)

POSITION OF LANDMASSES IN RELATION TO THE NORTH AND SOUTH POLES IN THE LATE SILURIAN, 420 million years ago

83

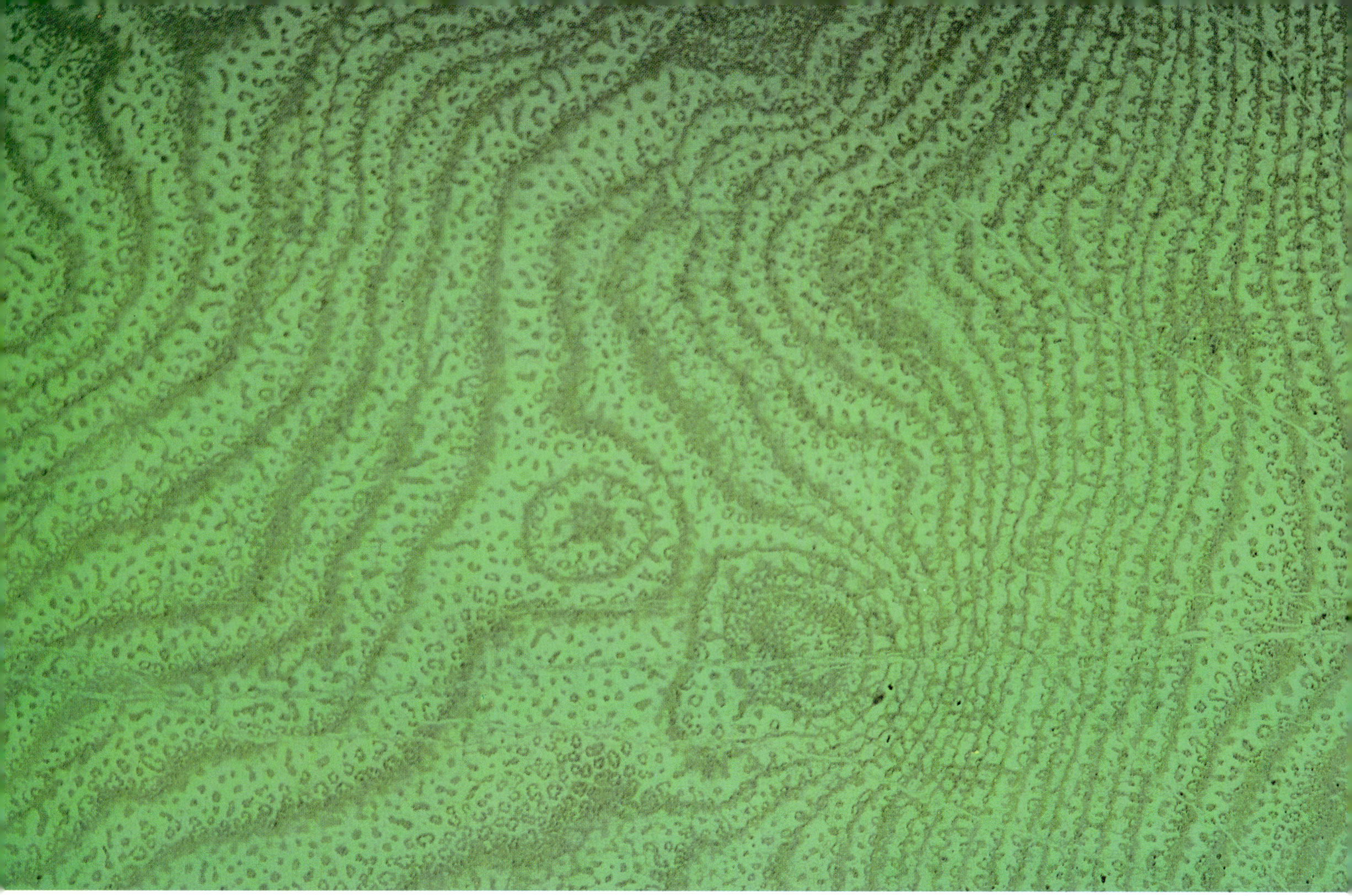
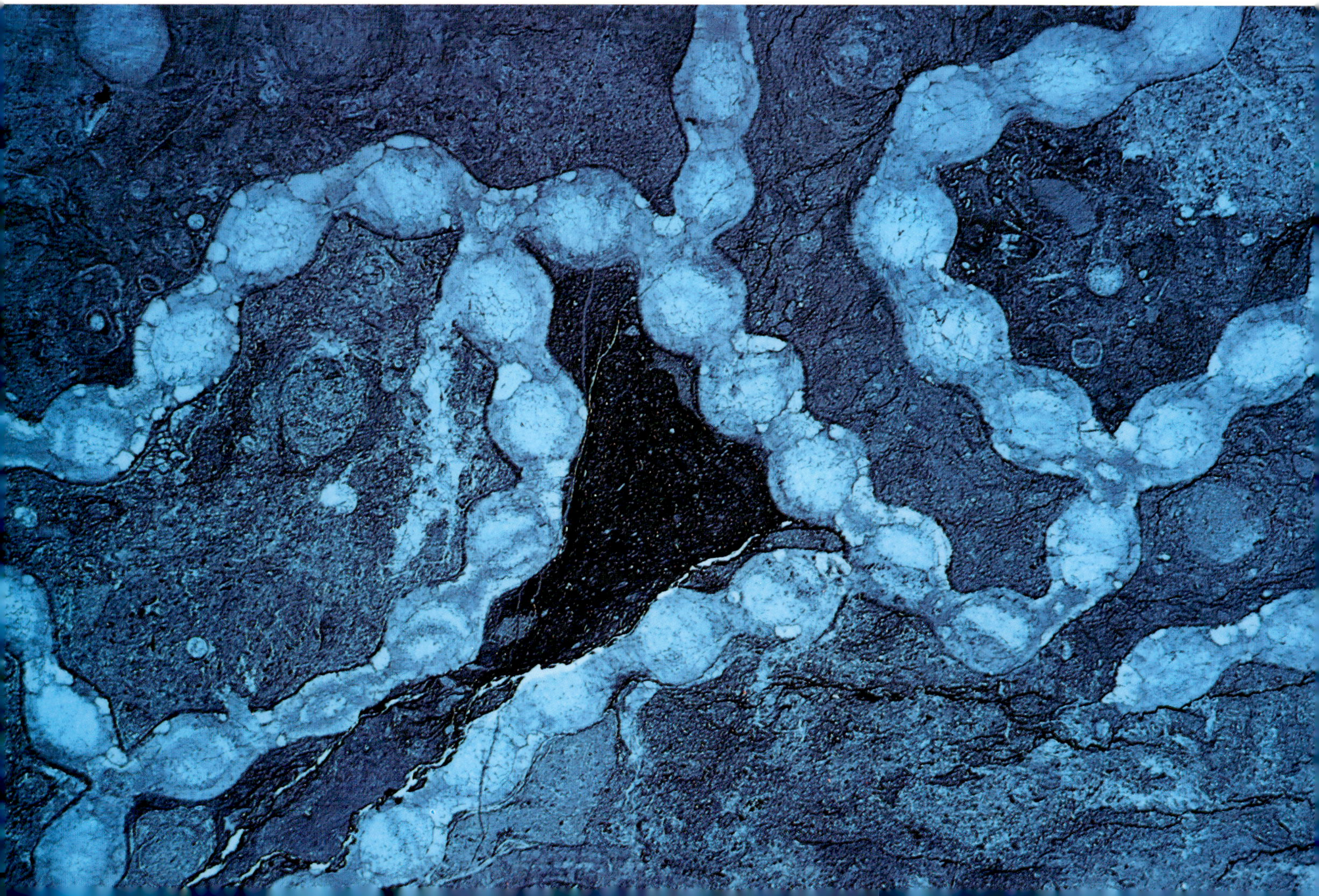

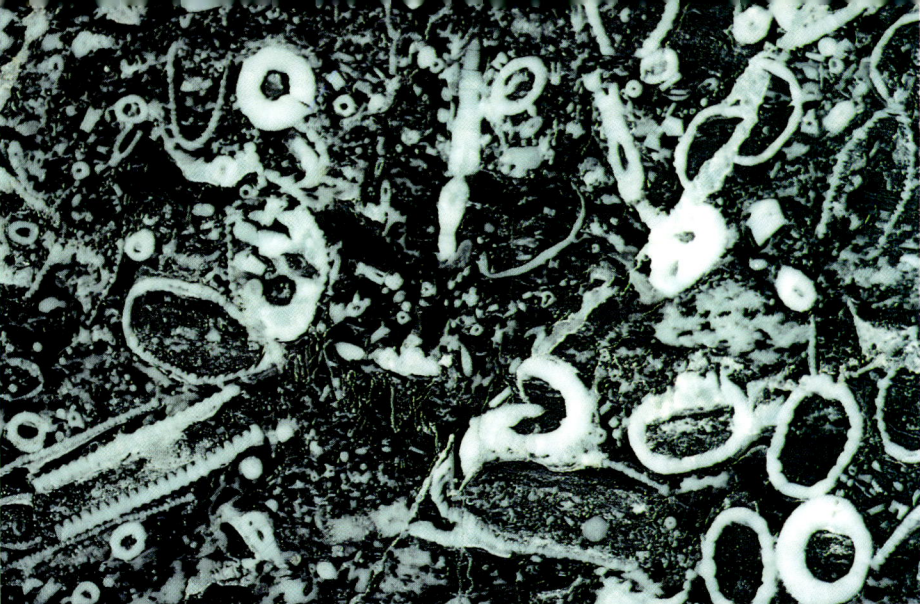

CRINOIDS

Crinoids are primitively stalked Echinoderms with long arms, and they typically lack respiratory structures. Some more advanced forms have lost their stalks and become secondarily free. The group was very diverse and important in Palaeozoic faunas, and remains of Crinoids contribute substantially to ancient limestones due to their preference for living in monodominant ''Crinoid gardens''.

Crinoids are rarely preserved complete. Stems are abundant, and great thicknesses of limestone may be made up almost exclusively of them. The body of the Crinoid is enclosed in a globular or cup-shaped plated structure known as the theca or calyx, which is usually broken up into its separate plates before fossilisation. The long plated arms (brachia) arise from the top of the theca.

The earliest Crinoids had rigid plated arms, and this type range from the Ordovician to the Permian. Mesozoic Crinoids developed flexible arms. Modern unstalked Crinoids live at all depths from sublittoral to abyssal. Stalked forms usually occur below 100 metres.

"Entrochal Marble", a limestone (which has been altered by heat or compression to become hard marble suitable for cutting into building stone) composed almost entirely of Crinoid stems. From Bathurst, New South Wales. (Specimen MA. Magn.X 1.3)

A complete Crinoid from England which shows the form of a whole organism. The stem terminates in a bulb (the plated calyx) and the arms are attached to the top of the bulb. (Specimen AMF. Magn.X 1.4)

A Crinoid head with plated tentacles. From Europe. (Specimen AMF. Magn.X 1.4)

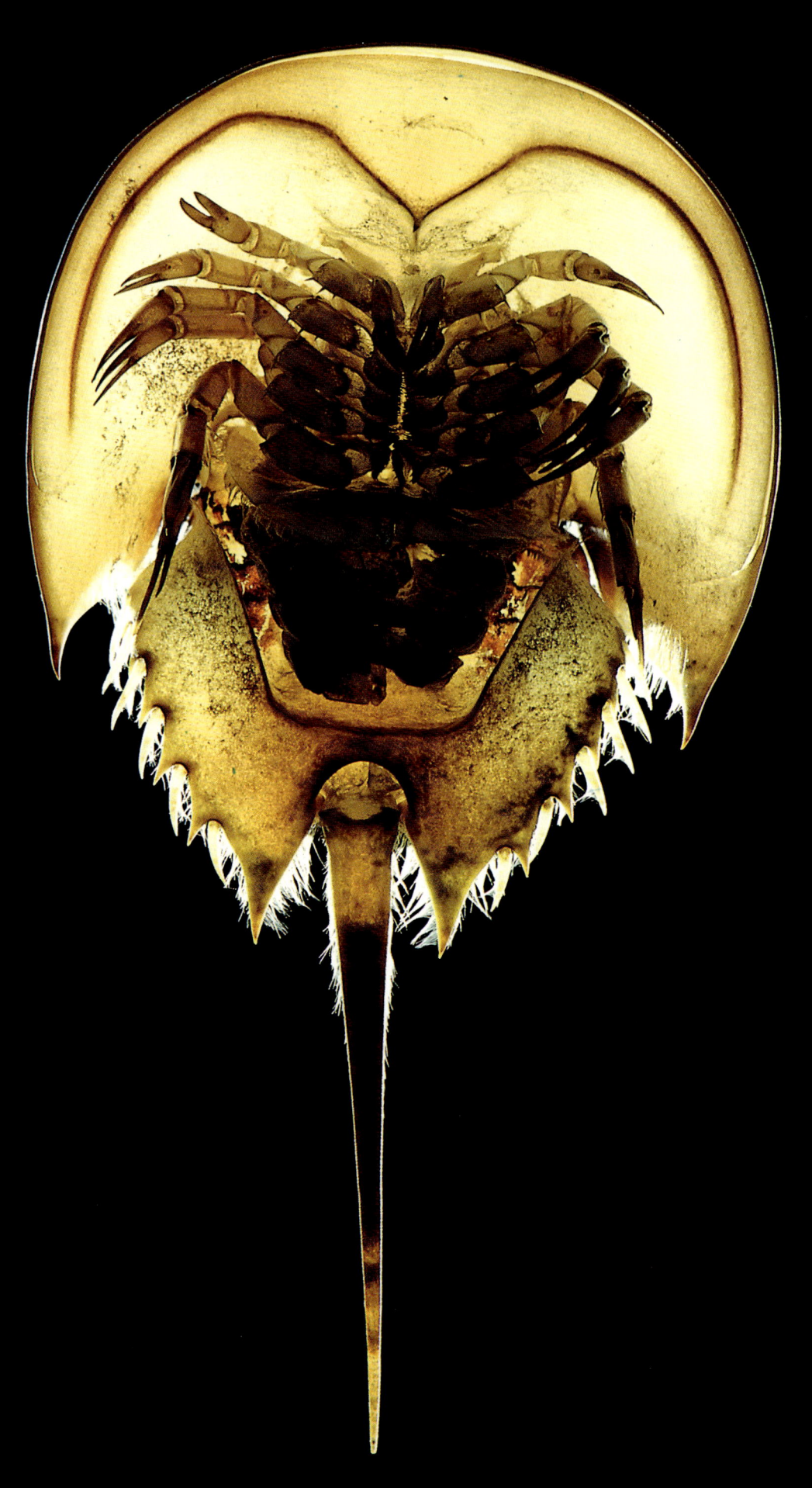

CHAPTER 2

LIFE VENTURES ONTO THE LAND

THE LATE SILURIAN PERIOD
FROM 420 TO 408 MILLION YEARS AGO
DURATION: 12 MILLION YEARS

The Late Silurian Period, that interval of geological time between 420 million years ago and the end of the Period some 408 million years ago, is of enormous significance in the evolution of life on our Earth. For the previous 3500 million years life had been restricted to the waters on the planet, and the land had remained an inhospitable wasteland. Now, at last, the time had come for change.

Oxygen levels in the atmosphere were now high enough to create an ozone layer of sufficient density to filter out harmful radiation to a point where life under air and not only under water became possible. The time had come to start the conquest of the land, and evolutionary processes were set in motion which would take a completely new direction. They would lead, ultimately, to Mammals and humans in the Animal Kingdom, and to modern vegetation in the Plant Kingdom.

So it was that towards the end of the Silurian Period the age of "living sea, dead land" came to an end. To the palette of hues of the dry and lifeless land, a new colour was added. A tide of green crept out of the water, tinging the swamps and outlining their margins. Green is the very colour of life, for it is the green plant pigment, chlorophyll, which enables plants to photosynthesise carbohydrates — the basis of food chains on which 99.9 per cent of Earth's inhabitants depend.

A green scum of Algae interacting with Bacteria and Fungi formed a zone, a halfway-house, between the water and the land. From the scum to the evolution of the first tiny land-plants tied to watery habitats was a small step for the plants concerned, but an enormously significant step in the evolution of the Plant and Animal Kingdoms which followed. It probably took another 20 million years of evolution in swamps before any invasion of real dry land was accomplished by plants.

So the land world of the Late Silurian and also the Early Devonian was largely an uninhabited desert of eroding rocks, offering neither food nor shelter for animals, and representing a realm beyond the capabilities of early plant life to invade it.

Limulus, a Horseshoe Crab. The Horseshoe Crabs are the only living representatives of the order XIPHOSURIDA of the Arthropod phylum. Fossil representatives range from the Ordovician to the Recent. The Eurypterids, whose range is from the Ordovician to the Permian, are closely related to Limulus. (Magn. X 1.7)

JIM FRAZIER

ANCIENT TRACE FOSSILS IN TUMBLAGOODA SANDSTONE

The red rocks of the Tumblagooda Sandstone in the river gorges in the Kalbarri National Park, Western Australia, contain an abundance of trace fossils.

Skolithos. These sand tubes, which are burrows made by animals, are abundant in the sandstone walls of Rainbow Valley in Kalbarri National Park. Tubes like these are often the only evidence that soft-bodied creatures like Annelid Worms existed at the time when the rocks were being formed, because the tissues of animals which had no hard parts hardly ever became fossilised.

Worm tubes and burrows (Skolithos) in red-beds in Tumblagooda Sandstone.

Worm tubes in Tumblagooda Sandstone.

Long tubes made by Worms which burrowed upwards as their habitat was progressively silted up.

THE FIRST FOOTPRINTS ON THE LAND

The Tumblagooda Sandstone in the Kalbarri National Park in Western Australia contains trackways made by Eurypterids walking on the land.

Eurypterid trackways on the rock platform.

Eurypterid tracks on a rock platform. Eurypterids were among the first creatures to venture onto the land. In the Late Silurian, when these tracks were made, the land was a dry and inhospitable desert, bare of vegetation except on the fringes of swamps and waterways.

Red sandstone cliffs of the Tumblagooda Formation, south from Nancy Beach, where Eurypterid tracks are etched into the rock platform.

ROGER HOCKING

ROGER HOCKING

Baragwanathia longifolia from the Late Silurian of Victoria. This fertile example of an ancient Clubmoss about 420 million years old has sporangia (seen as blobs on the bend of the frond) in the angles between the leaves and the stem. (Specimen JD. Magn.X 1.0)

In the Late Silurian, about 420 million years ago, the time came for life to venture onto the land. This was the dawning of a new age and the start of a miraculous transformation. The land world had waited through an immensity of time: it was a time-worn, inanimate realm governed by physical and chemical laws. Now as life was breathed into it, the land would change from that silent, sterile world to a green and lovely place alive with the sound of birds and the wind in the trees — to our world of the present day, and all in a time span which is so brief when it is compared with all the time that had gone before.

When early vascular plants became established on the fringes of the land they created a new zone and changed the environments on the edges of the water, in wetlands and in swamps. With the high sea levels of the times, and a climate that was hot and humid, there were abundant suitable areas for them to spread. The new environments imposed pressures to evolve, and pushed the boundaries of their habitats ever outwards. The success of plants in adapting to the new set of circumstances enticed animals to move onto the land to occupy the new niches which were being created. The race to inhabit the land world had started.

Plant life fringing the water created soil from sand, retained moisture and slowed erosion. There is no firm evidence of what sorts of plants were the first to make the transition from water to land. From a green algal ancestor to the first recorded land-plant is quite a jump. It seems likely that symbiosis between Algae, Fungi and Bacteria, and later between primitive vascular plants and Fungi and Bacteria, played a significant role in establishing life on the land. Some of the earliest vascular plants (known in detail as petrifactions in the Rhynie Flora in Scotland) have symbiotic Fungi (Mycorrhiza) in their tissues, showing that this important relationship which supplies nitrogen to the plant is a fundamental and primaeval one.

Eurypterids, Arthropods which looked somewhat like large Scorpions, were among the first animals to take to the land, and their tracks in Silurian sediments are the first footprints known to be preserved on land. Among Invertebrate phyla, only the Arthropods made the transition from water to a land-dwelling, air-breathing habit in significant numbers; in fact, there are terrestrial members of nearly all the classes of this phylum. In contrast, Earthworms and Leeches are the only members of the Annelid phylum to make the transition, and snails and slugs are the only Gastropod Molluscs to adapt to life on the land.

The peculiarities of organisation and structure of the Arthropod body suited this group of animals for terrestrial life. Their chitinous exoskeleton protected them against drying out by enclosing their delicate tissues in a microclimate. Chitin strengthened their limbs, enabling them to support their weight on the land, and their breathing systems readily adapted to taking oxygen from the air instead of from the water.

AUSTRALIA IN THE LATE SILURIAN PERIOD

In the Late Silurian, seas encroached on the western margin of the continent, and carbonates and other sedimentary deposits accumulated in the new shallow marine environments. The presence of evaporites, formed by evaporation of shallow embayments, provides evidence that there were times of aridity. As well, Australia's situation between the Equator and 30° North at the time implies a hot climate, while the presence of abundant Coral reefs in Australian waters and in those of other tropical and subtropical regions confirms that seas were warm.

An outcrop of Chillagoe Limestone forms a striking feature of the landscape near Mungana in northern Queensland.

The formation of elongated islands and troughs offshore from a narrow shelf area on the eastern seaboard, which had been a feature of earlier parts of the Silurian Period, continued. Sediments which collected in sheltered waters between the islands and the margin of the continent were to form a continuous rock sequence of Silurian to Devonian age. Those laid down in Late Silurian and Early Devonian contain the *Baragwanathia* Flora of Victoria, one of the most important and remarkable early land floras found anywhere in the world. The Late Silurian part of the sedimentary sequence is dated by the identification of a specific Graptolite which occurs with the *Baragwanathia* on the same faces of specimens.

Baragwanathia is a surprisingly highly organised plant for such an ancient horizon. It is a lycopod, or Clubmoss, and in its structure and its reproduction by spores produced in sporangia at the bases of some of its elongated leaves it resembles its living descendant, *Lycopodium squarrosum*.

Lycopods are descended from Zosterophylls, and some of these ancestral plants occur with the *Baragwanathia*. Other simple ancestral plants, Rhyniophytes — which are ancestral to Ferns, Horsetails and all the different kinds of Seed Plants — are also present in the *Baragwanathia* Flora.

NEW ZEALAND IN THE LATE SILURIAN PERIOD

No rocks of Silurian age have been recognised in New Zealand. Any sediments that might have accumulated in the basins on the outer edge of Gondwana, adjacent to Antarctica, would have been lost by erosion or obscured by subsequent geological events.

OVERLEAF:

THE LATE SILURIAN PERIOD
About 420 million years ago

Eurypterids, relatives of Horseshoe Crabs and up to a metre in length, were among the first creatures to walk on dry land, about 420 million years ago. Early land-plants created a fringe area on water margins, offering food and shelter and enticing Arthropods to adapt to life on the land. Three types of early land-plants are illustrated: at the right of the picture are *Cooksonia*, the first ancestral land-plants, with globular sporangia at the tips of their forking branches; in the foreground and middle right are Zosterophylls, with lateral sporangia, which are ancestors of Clubmosses (including *Baragwanathia*, which appears in the Fossil Record in Australia at about this time, and the plants around the Eurypterid, with forking stems and elongate terminal sporangia, are Rhyniophytes, ancestral to Ferns, Horsetails and all seed-bearing plants.

FIRST LIFE ON THE LAND

CHAPTER 3

LIFE IN THE EARLY SWAMPS

THE DEVONIAN PERIOD
FROM 408 TO 360 MILLION YEARS AGO
DURATION: 48 MILLION YEARS

The Devonian Period saw increasing adaptation of plants and animals to life on the land, though ties with the water were still binding. Only the swamps and water margins were vegetated and inhabited, with most of the dry land remaining bare and inhospitable.

For the first half of the Period plants were small and low-growing, reproducing by spores. The ability to produce supporting and conducting tissues in sufficient quantity to allow them to grow into trees was acquired by about the Mid Devonian, and by the end of the Period some plants had reached the stage of reproducing by seeds. The Giant Clubmosses and Giant Horsetails then characterised the flora, and swamps were densely vegetated and busy with animal life. The Arthropods increasingly made the transition to a land life, and fossil Spiders, Mites, Millipedes and Scorpions from that time, are known. The Lungfish bridged the worlds of water and land, and early Amphibians hunted in the swamps.

Fish and other life in the warm seas were diverse and abundant, and Corals and related organisms formed vast reefs. Fossil reefs at the northern margin of the Canning Basin in Western Australia, in particular, show the richness of the marine life.

Conodonts are microscopic phosphatic organs believed to be the tooth-plates or parts of the filter mechanism of Worm-like creatures, only one of which has been found as a fossil. (The animals were soft-bodied, with no tissues suitable for preservation.) Conodonts are abundant fossils, ranging from the Cambrian to the Triassic, and are useful for correlating rock strata, as they evolved and changed rapidly and different forms characterise different time zones.

Lepidosigillaria yalwalensis. A portion of the trunk of a Giant Clubmoss, patterned by leaf base scars. From the Mid Devonian, about 380 million years ago, when plants of this sort were the dominant trees in swamp forests. (Specimen MMF. Magn.X 1.5)

An Early Devonian Coral from Tarago, south of Goulburn, New South Wales. (Specimen MMF. Magn.X 1.0)

Landmasses of the Northern Hemisphere comprised a number of separate blocks during the Devonian Period. Gondwana lay in the Southern Hemisphere with only Australia, on its outer edge, straddling the Equator. World climates were warm to hot and there was no polar ice. Evaporites occurring abundantly worldwide indicate that aridity was a feature of climates. The abundant swamp-life of the Late Devonian resulted in deposits of organically-rich sediments. Important source rocks for the Texas, Canadian and Volga–Ural (USSR) oilfields formed from these Devonian swamps.

AUSTRALIA IN THE DEVONIAN PERIOD

Australia was situated between 15° North and 15° South in Early to Mid Devonian times and was the most tropical part of Gondwana. However, by the Late Devonian it had slipped further southwards to lie between 30° and 45° South, and was slightly differently orientated. Thus, up to the Mid Devonian the climate must have been hot and later it would have been somewhat cooler.

Volcanic activity on the eastern flank of the Australian continent continued to add new stabilised land and by the end of the Period the eastern margin approximated that of today. In the eastern segment of the continent during the Early Devonian, sediments accumulated on a large delta in the Adavale Basin, forming the reservoir rocks for the oldest known hydrocarbons found in eastern Australia — the Gilmour Natural Gas Field. At the same time deep ocean troughs near Melbourne and Bathurst were continuing to accumulate sediments, while shallow sea areas in eastern Victoria and near Canberra supported rich reef growth. A rich and diverse shallow-water fauna populated a shelf extending northwards through the New England area to northern Queensland.

During the Early Devonian the eastern and western seaboards were again linked by a wide ribbon of epicontinental sea — a successor to the Larapintine Sea — which provided a wealth of shallow-water environments. A major incursion into the Canning Basin formed the western part of this sea. In the middle of the Period an orogeny affected the eastern half of the Australian landmass, interrupting the eastern part of the shallow seaway and causing seas to retreat eastwards. With their retreat, more land was exposed on the eastern margin of the continent. Concurrently, elevation of Central Australia resulted in a westward retreat of the rest of the epicontinental seaway. By the Late Devonian all that remained was a small incursion into the western sector of the Canning Basin, and at that time there were also similar incursions into the Carnarvon and Bonaparte basins. (The west coast basins were subjected to regional subsidence from the Mid Devonian onwards.)

Erosion of the elevated Centre resulted in the production of sediments which banked up against the emerging mountains of the eastern sector, where the Tabberabberan Orogeny was creating a north-south trending mountainous region. This region of high ground formed a watershed for eastward-draining river systems, which spread large bodies of sediment. Freshwater Fish were abundant in the rivers.

In the Late Devonian, desert conditions were a feature of climates worldwide, and Australia was no exception. Parts of the Canning Basin region in Western Australia became desert after the marine incursions of the earlier part of the Period had retreated. The presence of widespread wind-blown sediments in many areas of the continent indicates that desert conditions were not uncommon. However, towards the end of the Period aridity decreased and a hot and humid climate prevailed.

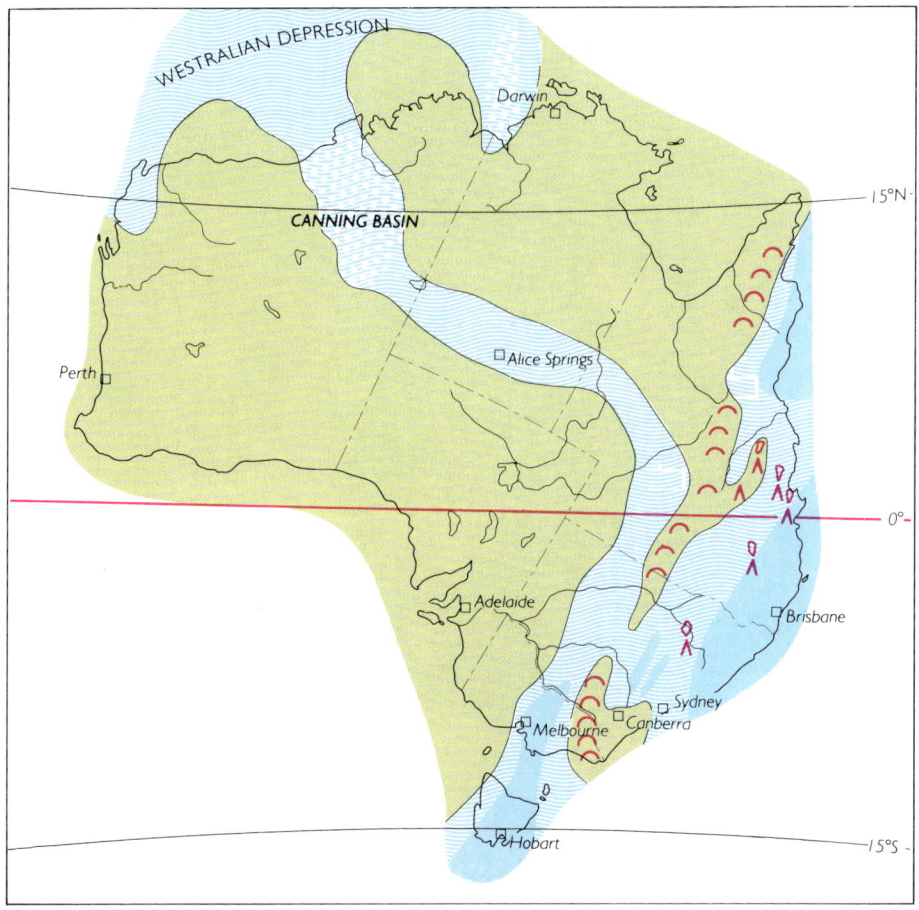

PALAEOGEOGRAPHY OF THE EARLY TO MID DEVONIAN
400 to 380 million years ago

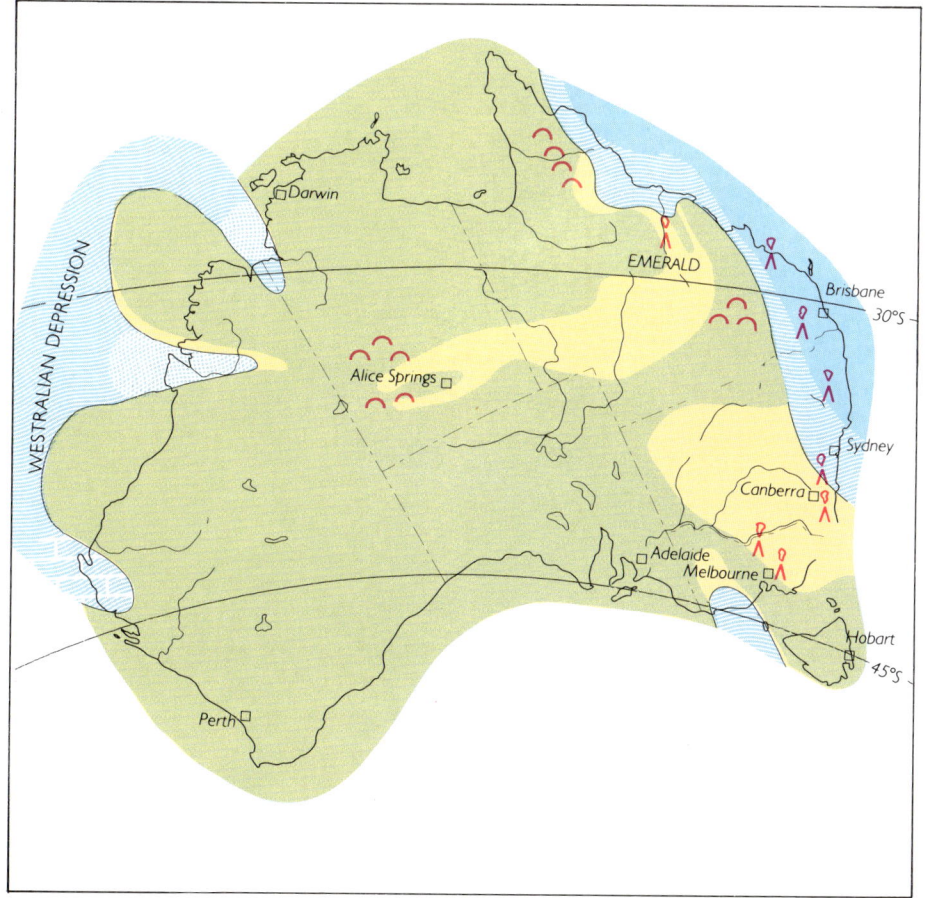

PALAEOGEOGRAPHY OF THE LATE DEVONIAN
365 million years ago

Fertile branches of an early gymnospermous Seed Plant, Barinophyton obscurum, from the Late Devonian at Genoa River in New South Wales. (Specimen MMF. Magn.X 1.1)

OVERLEAF:

LIFE IN A DEVONIAN SWAMP
About 380 million years ago

Land life was confined to swamps and water margins at this time, with the dry land, hillsides and mountains remaining uninhabited. Large Amphibians were the only land Vertebrates. Reed-like Horsetails with segmented stems, terminal cones and circlets of leaves are in the foreground, while large tree-growing Horsetails constructed on the same regular branching pattern are at the middle right. The large trees at left and right, with a bark pattern of regular rhombic leaf-base scars, are *Leptophloeum australe*, a Giant Clubmoss. The herbaceous Clubmosses in the right foreground resemble the older *Baragwanathia*.

97

LIFE IN A DEVONIAN SWAMP

RECEPTACULITIDS

The Receptaculitids are a group of Ordovician to Devonian "problematica" with such beautiful and complex structure that they are known as "Sunflower Corals". The consensus of opinion is that they are not related to Corals, nor are they Sponges. They are usually classified as a family within the order Dasyclydales of the Green Algae.

Receptaculitids are semi-globular or pear-shaped organisms from 1 centimetre to 30 centimetres in diameter. They have walls consisting of two layers connected by spindle-shaped pillars. There are two genera of Receptaculitids: *Receptaculites* and *Ischadites*.

In *Receptaculites* the outer and inner walls are composed of rhomboidal plates which lie exactly opposite each other, connected by the pillars. The plates are arranged in an ascending spiral (a feature clearly seen in the photographs of the external surface of the organisms). The outer plates are supported by four projections which form a cross at the top of each pillar.

Ischadites lacks plates on the inner wall (which is an undivided layer) and the outer plate support is three-pronged. The outer plates are spirally arranged as in *Receptaculites*.

Receptaculitids are environmentally significant, as they are found only in rocks which were formed in low latitudes (up to 20° North and South of the palaeo-Equator). They lived in shallow marine environments in warm seas.

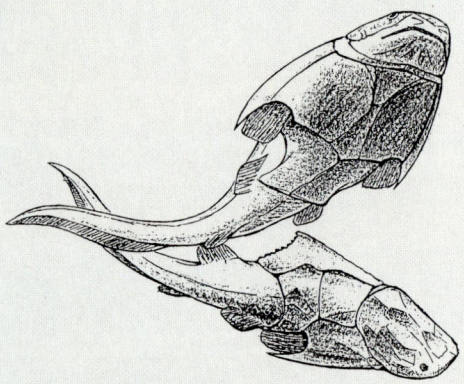

PLACODERM FISH
after Alex Ritchie, with permission

Receptaculites. A section through a petrified specimen which shows the structure of the wall with pillars between the inner and outer layers. From Devonian strata at Taemas, New South Wales. (Specimen AMF Magn. X 1.1)

Ischadites, showing the outer surface with some of its plates missing, thus exposing the pillars which connected them to the inner plates. (Magn. X 4.0)

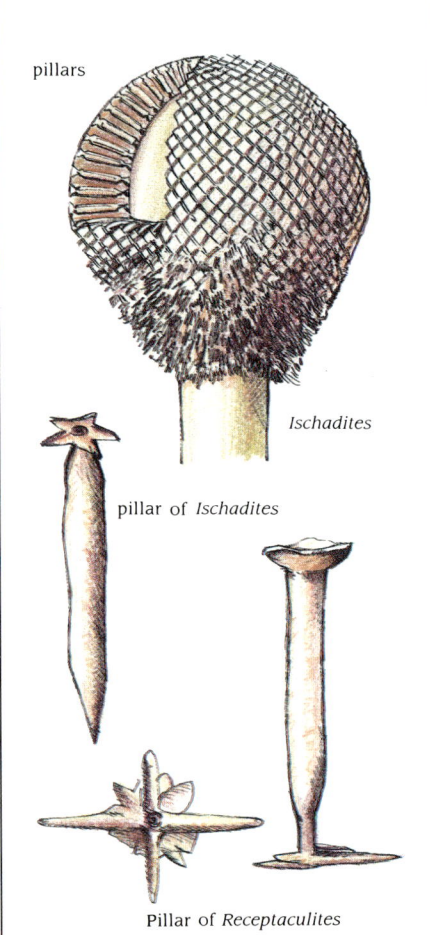

pillars

Ischadites

pillar of *Ischadites*

Pillar of *Receptaculites*

RECEPTACULITIDS
After Byrnes, 1968

Ischadites, showing the surface pattern made by the plates which form the outer wall of the organism. From the Mid Devonian Garra Formation of New South Wales. (Specimens AMF. Magn. X 2.2)

Ischadites, with pillars lying free at the top of the picture where the specimen is broken. The pattern of plates forming the outer surface is seen below (Magn. X 2.2)

DEVONIAN REEFS OF WESTERN AUSTRALIA

The classic face of the reef at Windjana Gorge.

The reef forms a prominent feature of the landscape.

Lloyd Hill, an atoll, from the air.

Early Devonian plants were herbaceous and low-growing, and were still tied to watery habitats. The *Baragwanathia* Flora of Victoria, which originated in the Late Silurian, persists into the Early Devonian. It comprises Zosterophylls (which are ancestral to Clubmosses), *Baragwanathia* (which is a Clubmoss, and similar to species of *Lycopodium* alive today) and Rhyniophytes (which are ancestral to Ferns, Horsetails and all kinds of Seed Plants). Some Rhyniophytes present in the Early Devonian are of an advanced type called Trimerophytes, which are intermediate forms ancestral to the Seed Plants.

By Mid Devonian times land-plants had acquired the ability to grow into large trees. The age of the Giant Clubmosses and Giant Horsetails had dawned, and swamps now supported forests. In Australia, *Lepidosigillaria* appears in the Fossil Record.

Leptophloeum australe, a giant Clubmoss with a rhombic leaf-base patterning on the trunks, is the characteristic plant of the Late Devonian. By the end of the Period species of *Lepidodendron* Clubmosses appear. Early Gymnosperms (Seed Plants) are also found, though they are still rare. The evolution of seeds was a significant advance, suiting plants to the dry land.

For the Animal Kingdom, the Devonian Period was the Age Of Fishes. Marine and freshwater Fish were abundant, including Lungfish. Placoderms with armoured heads and bodies from central and south-eastern Australia show affinity with those of southern China. This distribution is evidence that large areas of land which would become parts of Asia were at this time attached to Gondwana.

The discovery of Labyrinthodont Amphibian footprints at Genoa River in Victoria, and of a jaw from the Clognan Shale located south-west of Forbes in New South Wales, are interesting and significant finds. The only other Tetrapods known from Devonian beds are from Greenland. The Late Devonian Amphibians signal the start of the conquest of the land by Vertebrates.

There was abundant Invertebrate marine life. Ammonites first appear in the Devonian. Their coiled shells are of types similar to those found in Europe, North Africa and North America. Receptaculitids, organisms of complex structure and hollow spherical form, are found in abundance as fossils in the Early Devonian Garra Formation of central western New South Wales. Brachiopods, Corals, Bivalves, Gastropods, Coelenterates, Sponges, Echinoderms and Conodonts were abundant in the rich faunas. The diversity of Trilobites was declining rapidly and the Graptolites had almost disappeared from the Fossil Record by the Devonian.

Late in the Period, massive reefs formed by Stromatoporoids, Algae and Corals flourished in the Canning Basin. Today these reefs form a broken line of ranges over a distance of more than 300 kilometres.

Rugose and tabulate Corals were abundant in the fore-reef and back-reef deposits.

Gogo Fish of the Gogo Formation near Fitzroy Crossing in Western Australia are preserved where they lived near the reefs. They comprise various armoured Fish (Placoderms), early ray-fin Fish and early Lungfish. On the world scene early Sharks were abundant at this time, and some are known from Bunga Head near Eden in New South Wales.

NEW ZEALAND IN THE DEVONIAN PERIOD

Sediments which accumulated in a shallow marine basin on the outer edge of Gondwana formed the Devonian rocks which are found in New Zealand. They are of limited extent. The Baton Formation, north-west of Nelson, and

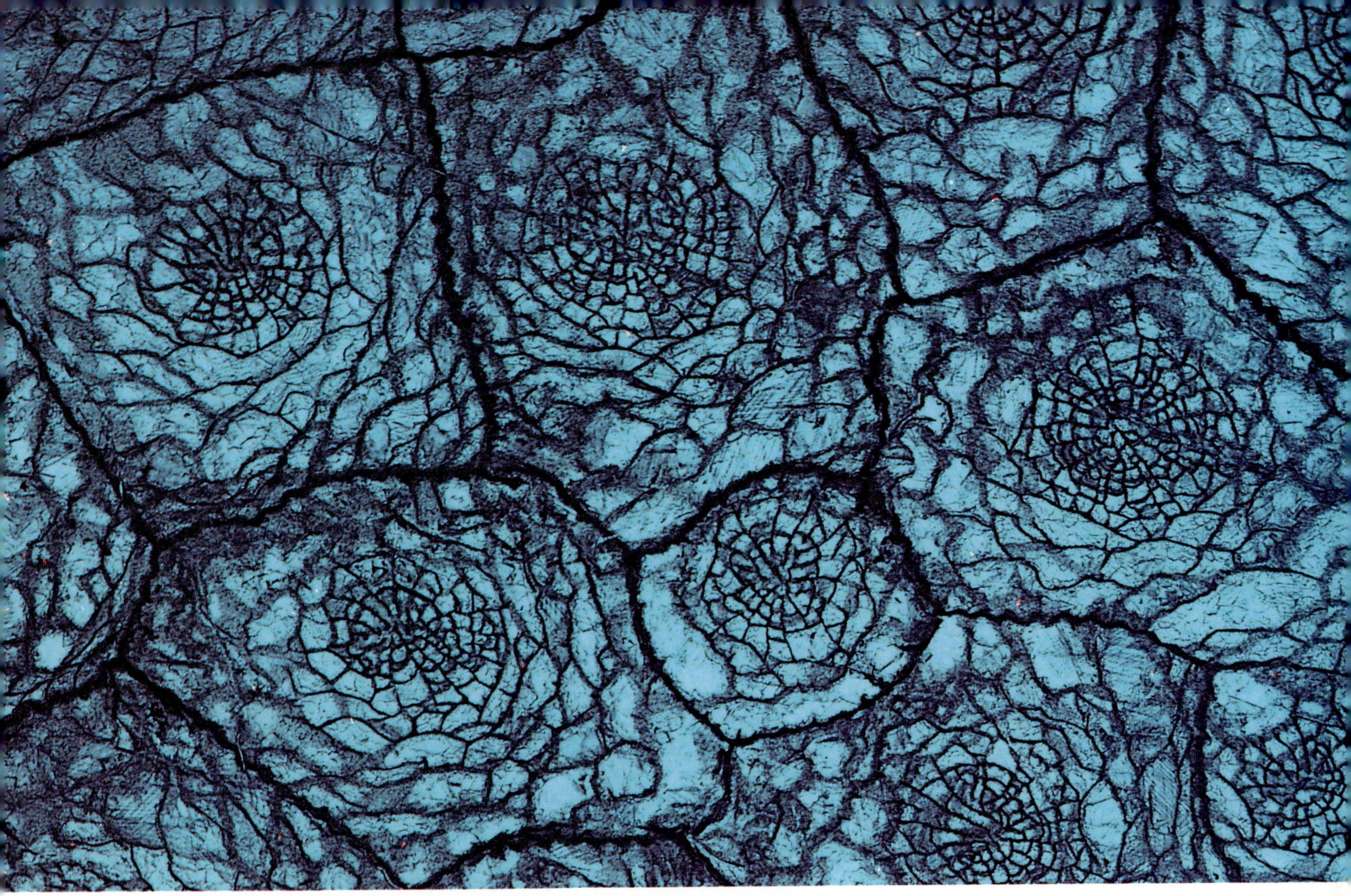

A thin section of *Spongophyllum giganteum*, a Coral from the Mid Devonian at Tamworth, New South Wales. (Specimen AMF.)

Phillipsastraea maculata, a Coral from the Devonian at Attunga in New South Wales. (Specimen AMF. Magn.X 1.2)

A piece of Coral, *Favosites*, from Wellington Caves, New South Wales. Age Early Devonian, about 400 million years. (Specimen MMF. Magn.X 1.4)

the Reefton Mudstones and Limestone, in South Island, have been identified as Devonian by their fauna. Many genera found in New Zealand occur in rocks of the same age in Australia. Receptaculitids, Corals, Brachiopods, Bivalve and Gastropod Molluscs, and Trilobites are present.

Devonian Corals in the region indicate that seas were warm. The types of sediment in which the fossils occur suggest that deposition occurred in a low-energy environment, which implies that the adjacent part of Gondwana was an area of low relief.

CHAPTER 4

MOVING DOWN UNDER

AN EPIC JOURNEY TOWARDS THE SOUTH POLE
THE CARBONIFEROUS PERIOD
FROM 360 TO 286 MILLION YEARS AGO
DURATION: 74 MILLION YEARS

During the 74 million years of the Carboniferous Period major rearrangements of the position of lands on the surface of the globe occurred. The opening up of the Tethys Sea swung the southern supercontinent of Gondwana southwards, sending Australia, which was on its outer edge, into South Polar regions. With an ice age approaching, Australia moved into increasingly cooler climatic zones. When the movement was completed, the landmass lay in polar latitudes and half of its area was covered by a continental ice sheet.

Life was still confined essentially to the swamps and areas of high water-table during Carboniferous times. Hillsides and dry areas were not vegetated. Vast regions of the globe must have had landscapes of bare rock, gravel and sand like those of the ancient world before the land was inhabited.

The swamps of equatorial regions continued to support luxuriant forests dominated by Giant Clubmosses and Giant Horsetails. Insects were abundant, and they and many other Arthropods had adapted to life on the land. Vertebrate land animals were represented by Lungfish, Amphibians and early Reptiles.

But in Australia, as the climate deteriorated and the ice age intensified, the flora and fauna became impoverished. By late in the Period only a low-growing tundra vegetation of pro-Gymnosperms survived around the ice sheets, while the cold seas supported low diversity, high latitude assemblages of marine creatures.

No rocks of Carboniferous age have been found in New Zealand.

Rugose Corals, also called Horn Corals, are common in Palaeozoic rocks all over the world. They became extinct early in the Triassic. This specimen from Jervis Bay in New South Wales is approximately 300 million years old.
(Specimen MMF, Magn.X 4.1)

Notorhacopteris ovata, a pro-Gymnosperm which grew on the tundra. From Stroud, New South Wales. Age about 300 million years. (Specimen MMF. Magn.X 1.0)

At the start of the Carboniferous much of the world's land lay in tropical and subtropical zones. The ancestral Giant Clubmoss Flora, which had become established in swamps in the Late Devonian, evolved and flourished in all the warm lands. The progressive opening of the Tethys Sea, separating Gondwana from the lands of the Northern Hemisphere, brought Australia and the southern parts of South America, Africa and India into high latitudes.

World climates were warm in the Early Carboniferous. With the changed circulation patterns that resulted from the rearranging of areas of land and sea, lands in higher latitudes were affected by a progressively increasing temperature gradient from the Equator to the Poles. An ice age was approaching, and a global drop in sea level in the Mid Carboniferous indicates the establishment of permanent polar ice.

Climates in the Euramerican lands which remained in equatorial regions continued to be hot and wet, and were little affected by the polar glaciation. The Giant Clubmoss Flora thrived under these conditions and contained Seed-ferns, Ferns and pro-Gymnosperms as well as the Clubmosses and Horsetails. The coal deposits of the Northern Hemisphere and the oilfields of western Texas and the western United States of America bear testimony to the luxuriance of this swamp vegetation.

The tropical swamps teemed with life. Insects were abundant, notably Dragonflies with a wingspan of over a metre, Mayflies and Beetles. Cockroaches and Centipedes were also common. The dominant land Vertebrates were Amphibians, but there were early Reptiles as well.

The South Pole was situated on Gondwana, and consequently the effects of glaciation and the formation of a permanent ice cap were profound on the southernmost lands. (The reverse was to be the case in the most recent Pleistocene ice age when northern lands bore the brunt of the ice, being clustered round the North Pole, while scattered southern lands were little affected in comparison.) It is likely that climatic fluctuations characterised the Late Carboniferous, as glacial and interglacial phases alternated.

The ancestral Giant Clubmoss Flora did not thrive in the ever-cooler southern lands and by late in the Period, with the advance of the ice sheets, it had been largely replaced by tundra vegetation.

AUSTRALIA IN THE CARBONIFEROUS PERIOD

At the beginning of Carboniferous times Australia lay in much the same latitudes as in the Devonian, but it had started to rotate slightly and was about to begin its epic journey to the South Pole. During the 74 million years of the Period it travelled from equatorial regions into high latitudes, and by the Late Carboniferous the north-south axis was at right angles to that of the present day. In its high latitude situation it was subjected to long months of winter darkness, intensifying the cold. The appearance of glaciers on high ground heralded the formation of the permanent ice cap, which was to cover more than half of Gondwana.

The geological evolution of the continent also escalated during Carboniferous times. Interaction between the stabilised land of the continent and the proto-Pacific Plate resulted in major tectonic activity in the Yarrol and New England provinces. Volcanic activity, deformation of rock strata, intrusion of igneous rocks and the raising of land above the sea increased the area of stabilised land. Along the new margin, fractured depressions developed and their remains now form the Sydney, Galilee and Bowen basins. However, in their original state, these depressions were

Lepidodendron and Stigmaria. The fine bark pattern (left and centre) of a Giant Clubmoss has an impression of part of a root (Stigmaria) with round spots overlying it. (Specimen MMF. Magn.X 1.2)

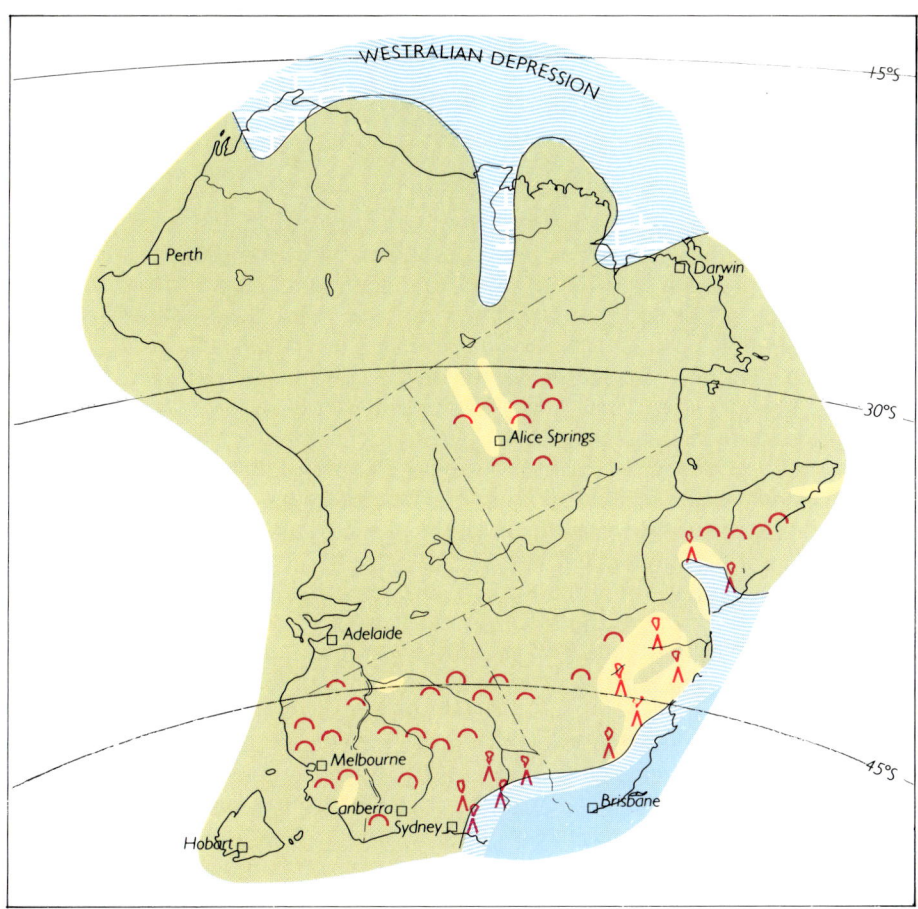

PALAEOGEOGRAPHY OF THE EARLY CARBONIFEROUS
345 million years ago

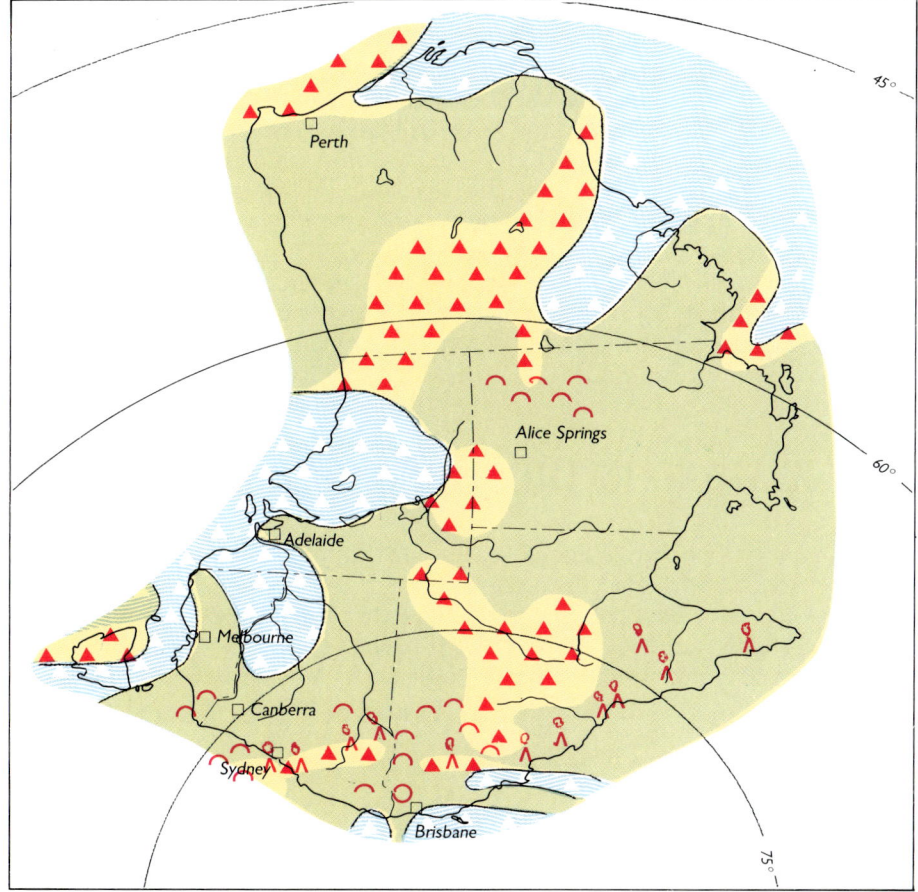

PALAEOGEOGRAPHY OF THE LATE CARBONIFEROUS TO EARLIEST PERMIAN
300 to 280 million years ago

Carboniferous Gastropods. The sea shells of 300 million years ago looked just like those we find in the intertidal zone today. (Specimen MMF. Magn.X 1.0)

CARBONIFEROUS SPORES AND POLLEN

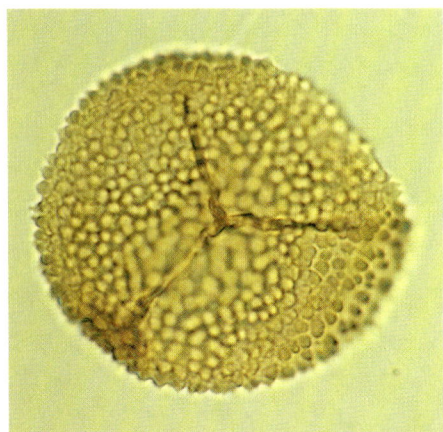

*Granulatisporites frustrulentus,
a microspore.*

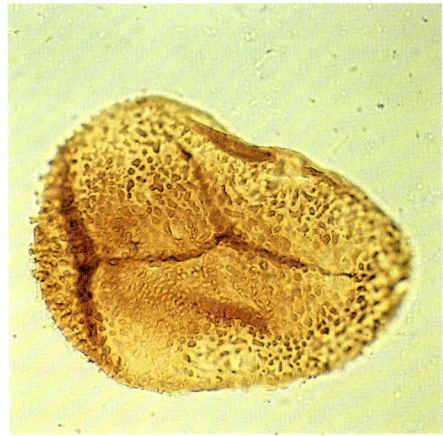

Spelaeotriletes ybertii, a microspore.

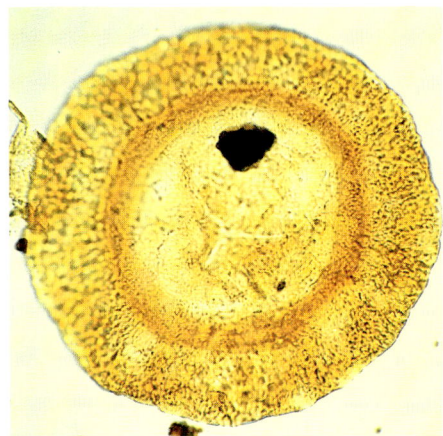

Plicatipollenites, a pollen grain.

trough-like complexes of localised rifts formed by tearing stresses generated by plate interaction. Volcanoes were associated with some of the rifts. Shallow seas flooded into the troughs, and sediments derived from the land and from the volcanoes gradually filled most of them. Those depressions which remained continued to subside, forming the basins.

Within a previously stabilised sector of the continent, in an area occupied today by north-western New South Wales and south-western Queensland, a zone of crustal weakness developed and a depression formed. It was to become the Cooper Basin: in the following Permian Period, swamp plants and abundant Algae in lakes would contribute kerogen to the sediments accumulating in the depression, resulting in the natural gas, light oils and condensates which now occur there.

Continued uplift in the Alice Springs Orogeny resulted in mountainous terrain in the area where the MacDonnell Ranges are today.

With the Australian continent moving down into South Polar latitudes and the approaching ice age causing a deterioration in climate, the meagre Giant Clubmoss Flora dwindled. As a result no coal was produced here, in contrast to the situation in the Northern Hemisphere. Tundra conditions prevailed on the margins of the ice and had to be endured for millions of years up to the Permian boundary and for some millions of years beyond it. An impoverished *Rhacopteris* Flora grew on the tundra.

An ice sheet covered more than half of Australia's surface during the ice age, and had a dramatic effect. It cleared the land of all living things, setting the stage for a great burst of evolution when the ice melted and new environments emerged.

Australia's southward movement into high latitudes and the accompanying ice age regime had as profound an effect on the fauna as on the flora. In the Early Carboniferous the cosmopolitan warm-water faunas were rich in Corals, Brachiopods, Echinoderms (especially Crinoids), Bryozoans, Molluscs and Foraminifera. Trilobites were still present, though declining. Fish of all types, including Sharks, were abundant. But as climatic cooling progressed, faunas became increasingly impoverished and only a few species were to survive. With the establishment of a high latitude biota, endemism increased.

The only Insect recorded from the Late Carboniferous in Australia occurs in glacial sediments in Tasmania. It is associated with fossilised Fairy Shrimps, and shows that some hardy creatures managed to exist in frozen regions, as is the case today.

NEW ZEALAND IN THE CARBONIFEROUS PERIOD

No rocks of Carboniferous age have been found in New Zealand. Their absence may be attributed to massive volcanic activity late in the Period, which would have obscured such sediments as might have been deposited — and probably were — in the offshore basins adjacent to Gondwana where New Zealand's foundations were being laid.

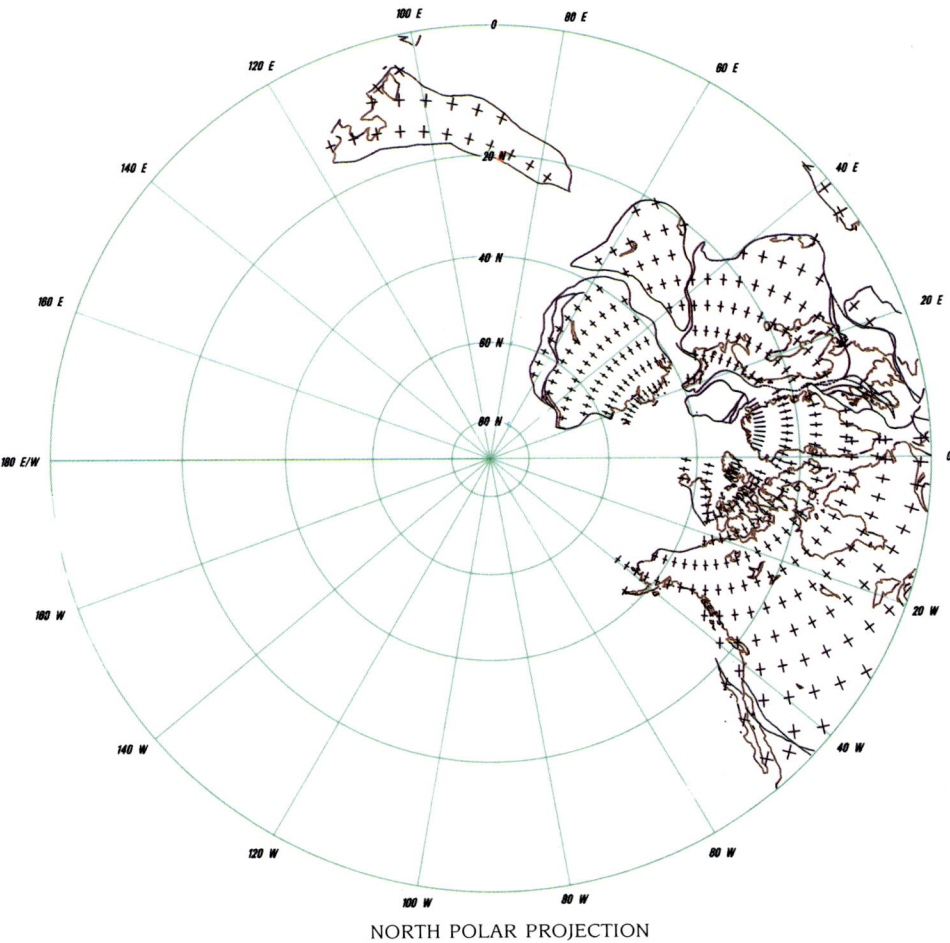

NORTH POLAR PROJECTION

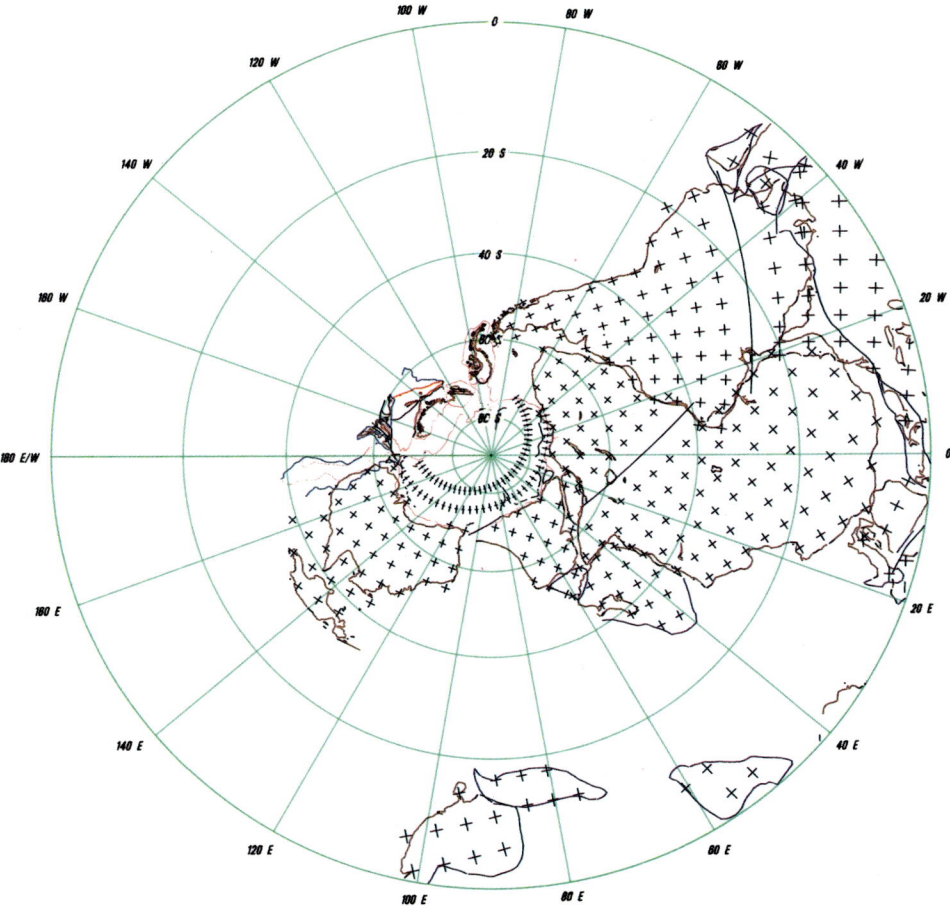

SOUTH POLAR PROJECTION

POSITION OF LANDMASSES
IN RELATION TO THE
NORTH AND SOUTH POLES
IN THE CARBONIFEROUS PERIOD,
300 million years ago

BOB TINGEY

The frozen high latitudes of Gondwana, under an ice cap during the Late Carboniferous to Early Permian ice age, resembled modern Antarctica.

OVERLEAF:

THE TUNDRA OF THE CARBONIFEROUS PERMIAN ICE AGE

About 290 million years ago

In the melt-water swamps round the ice sheets which covered much of Gondwana, a hardy cold-adapted flora of low-growing plants survived on the tundra. Rhacopterid Seed-ferns, peat Mosses, herbaceous Horsetails and Clubmosses formed a vegetation living close to the ground where they were protected from the icy winds.

THE TUNDRA OF THE CARBONIFEROUS PERMIAN ICE AGE

CHAPTER 5

COAL SWAMPS AT THE SOUTH POLE

THE PERMIAN PERIOD
FROM 286 TO 245 MILLION YEARS AGO
DURATION: 41 MILLION YEARS

At the conclusion of the Late Carboniferous to earliest Permian ice age, there was a new burst of evolution. With climatic conditions slowly improving to cool-temperate, the Glossopteris Flora diversified and spread all over Gondwana. Severe winters with months of polar darkness made conditions extreme at first, but they gave way to less rigorous conditions. Large inward-draining floodplains were a feature of landscapes, and swamps developed on them. The immense reserves of black coal mined in Australia and other Gondwanan lands today were formed from the organic remains of plants growing in these cool-temperate swamps.

A sequence of Permian marine sedimentary and volcanic rocks accumulated in a deep and sinking trough on the eastern margin of Gondwana adjacent to Antarctica. These rocks would later be raised up to become part of the New Zealand landmass. There they survive today as the most complete Permian sequence known anywhere in the world — up to 20,000 metres thick!

Shallow-water marine faunas showed the same burst of evolution after the ice age as did plants. Sea levels were high, so there was a wealth of shallow-water environments. At the end of the Period, a drop in sea level and the consequent contraction of these environments contributed to the large number of extinctions which characterise that time.

Labyrinthodont Amphibians were the dominant land Vertebrates during the Period and Reptiles were also abundant.

Vertebraria, the root of a Glossopterid, which has a segmented structure due to the inclusion of blocks of aerating tissue to assist it to function in waterlogged swamp environments. Fossils look like vertebral columns as a result of this specialised structure. The specimen shown is preserved in a lightweight kerosene shale from Joadja Creek in New South Wales. During World War II petrol was distilled from oil shales such as this. The extraction processes are costly and under normal circumstances production of petrol from such oil shales is not an economically viable proposition. (Specimen MMF. Magn.X 3.6)

A volcanic plug forms an impressive feature of the landscape near Clermont in Queensland. It is evidence of Permian volcanic activity in the area.

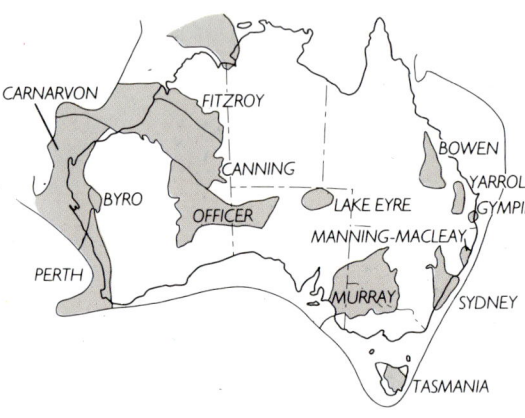

PERMIAN SEDIMENTARY BASINS

The Late Carboniferous to earliest Permian ice age was still intense at the 286-million-year time boundary between the two Periods. It is not certain whether there was a massive, continuous ice sheet on Gondwana like that which covers Antarctica today or whether there were a number of centres, changing through time. Evidence suggests that glaciation in South America and southern Africa is older than that in Australia, favouring the latter situation. In either case, it is likely that some areas, by reason of their topography, would have been refuges for plants and animals even when a large ice cap existed in their region. (The present-day ice cap in Antarctica is believed to have been in place for the last 15 million years. However, fossil wood preserved in sedimentary rocks of the region's Sirius Formation shows that forests grew there in the last 5 million years. This evidence implies that there was a local hiatus in the ice cap where cool-temperate conditions suitable for the growth of trees existed. Such seemingly anomalous conditions must surely have been features of the Carboniferous–Permian ice age as well.)

The presence in Gondwana of Early Permian glacial deposits interbedded with layers of coal which indicate temperate conditions is further evidence that forests and ice can co-exist. New Zealand and South America today have forests at the foot of glaciers.

The difference between conditions prevailing in the Euramerican landmass in equatorial latitudes and those in parts of Gondwana near the South Pole is well documented. There is evidence that parts of the warm-to-hot low latitude belt were affected by aridity during Permian times.

Major faunal changes had occurred globally by the end of the Permian. Brachiopods declined markedly and would not be influential again. Crinoids also declined dramatically and by the end of the Period nearly all the Palaeozoic stocks became extinct. Nautiloid Cephalopods continued to decline, while Ammonites became more diverse. Trilobites became extinct, as did Eurypterids, and Horseshoe Crabs declined. But there was a great increase in the abundance and diversity of Insects on the land.

AUSTRALIA IN THE PERMIAN PERIOD

Gondwana had been profoundly affected by the ice age, during which an ice sheet had covered half of its area — including half of the Australian landmass. As in other parts of the world, the high sea levels following the ice melt were to have significant consequences.

The glaciation of so much of the land was important in terms of fossil fuels. The debris left on the land surface on the retreat of the ice was subsequently reworked and deposited as reservoir rocks in the younger petroliferous basins.

After the ice melted in earliest Permian times, climates improved, though they remained cool to cold throughout the Early Permian, with glaciers on the eastern highlands. The dark winters must have been bitterly cold. As the Period progressed and warmer climates developed, a burst of evolution occurred, producing a rich and diverse flora in the coal swamps. (A parallel for these cool-temperate coal swamps is seen in the peat bogs, with their Birch and other trees, which characterised Scotland after the ice retreat in the most recent interglacial times.) Coal swamps were to persist in widespread areas until late in the Permian, under conditions which were at best warm-temperate. In the lakes associated with swampy areas, Algae were abundant and created a major source of kerogen from which natural gas, condensates and light oils are derived. The commercial deposits in the Cooper and Bowen basins, in particular, resulted from this time.

This seam of black coal at Blair Athol Mine in Queensland is the thickest one known in the world. The coal mined here, and most of the massive coal deposits mined elsewhere in Australia, was formed in the cool-temperate swamps which were a feature of Permian landscapes in Gondwana.

Eurydesma and a straight-shelled Nautiloid (centre). The five shells to the right of the Nautiloid are in growth position, with their heavy bases deeply impressed into the sediment. The distribution of weight, concentrating it in the base of the shells, resulted in the organisms being able to return to an upright position when knocked over by wave action in the high-energy environment in which they lived. Their niche was the rocky fringes of the cold seas in the Early Permian when the world was emerging from an ice age. (Specimen MMF. Magn.X 0.3)

Brachiopods of the Dielasmid type, which are almond-shaped and unlike the typical "Lamp Shell" varieties. (Specimens MMF. Magn.X 1.5)

The climate remained markedly seasonal throughout the Permian Period. Pronounced annual rings in the abundant petrified wood found in rocks of this age are partly evidence of this climatic fluctuation. It must be remembered, however, that the dormancy imposed by months of winter darkness was a contributing factor to seasonal growth patterns, and it is not possible to separate its effect from that produced by temperature changes.

In Australia, early in the Period, seas covered large areas of South Australia and Victoria. An arm of the sea on the Antarctica-Australia joint line separated the Tasmanian sector from the mainland. There were also large sea encroachments on the Western Australian coast and into the Canning Basin. Major changes in sea level as well as tectonic events controlled the distribution of sediments, which were accumulating in Permian basins.

There were apparently two major sea level rises during the Period. The first, due to melting of the ice cap, caused the flooding of the Carnarvon and Canning basins in the west of the continent, and of the Sydney and Bowen basins in the east. In the Mid Permian, the sea again invaded the Sydney and Bowen basins, in the last major episode in which the eastern margin of the continent was to be affected by a marine incursion. Local subsidence was the cause of this Mid Permian inundation. During the Late Permian the second global sea level rise occurred, but this time the eastern sector was not invaded by epicontinental seas. Instead run-off waters were impounded in the major basins, creating vast floodplains. Major petroleum reservoir and source rocks were created in eastern basins during this phase, and the black coal deposits mined today resulted from the swamp forests growing on the floodplains. The Cooper, Galilee, Bowen and Sydney basins were all influenced by the inward draining and silting regimes.

In the Canning Basin in Western Australia, and in the central zone of the Westralian Depression, sediments were accumulating. In the Bonaparte Gulf region, deltas supplied the sediments which form the reservoir rocks for the natural gas accumulations of that area.

During the Permian Period there was volcanic activity in the Sydney Basin and northwards on a line to Townsville. Rocks in the New England area were folded, intruded by granites, and uplifted in tectonic movements which resulted from plate interactions in the Yarrol–New England provinces. These areas were the last in which activity of the Tasman Orogen still persisted. Indeed, most of the Australian landmass was stabilised by then. The eastern Australian basins continued to sink as the weight of sediments in them increased.

In the west of the continent during Permian times, the Westralian Depression became increasingly pronounced and assumed its dog-leg shape. (It was to become a rift zone in the Mesozoic Era when the split-up of the supercontinent began.)

The distinctive vegetation that characterised the cool-temperate Gondwanan lands during the Permian Period is the *Glossopteris* Flora. The first Glossopterids to appear in Australia are found in Early Permian glacial deposits at Bacchus Marsh in Victoria. There *Gangamopteris* grew close to the glaciers which remained after the ice sheet had retreated. Rapid diversification saw the evolution of the order Glossopteridales. The plants of this group have an important place in the evolutionary story of the Plant Kingdom. They are probably ancestral to several groups of plants which appeared in the Fossil Record by the end of the Permian Period. They were adapted to cool climatic conditions and to growing in swampy ground. Their root systems were constructed to incorporate blocks of aerating tissue to overcome the problems of growing in saturated mud. By the end of the Period evolution within the order, along several lines, saw the efficient and

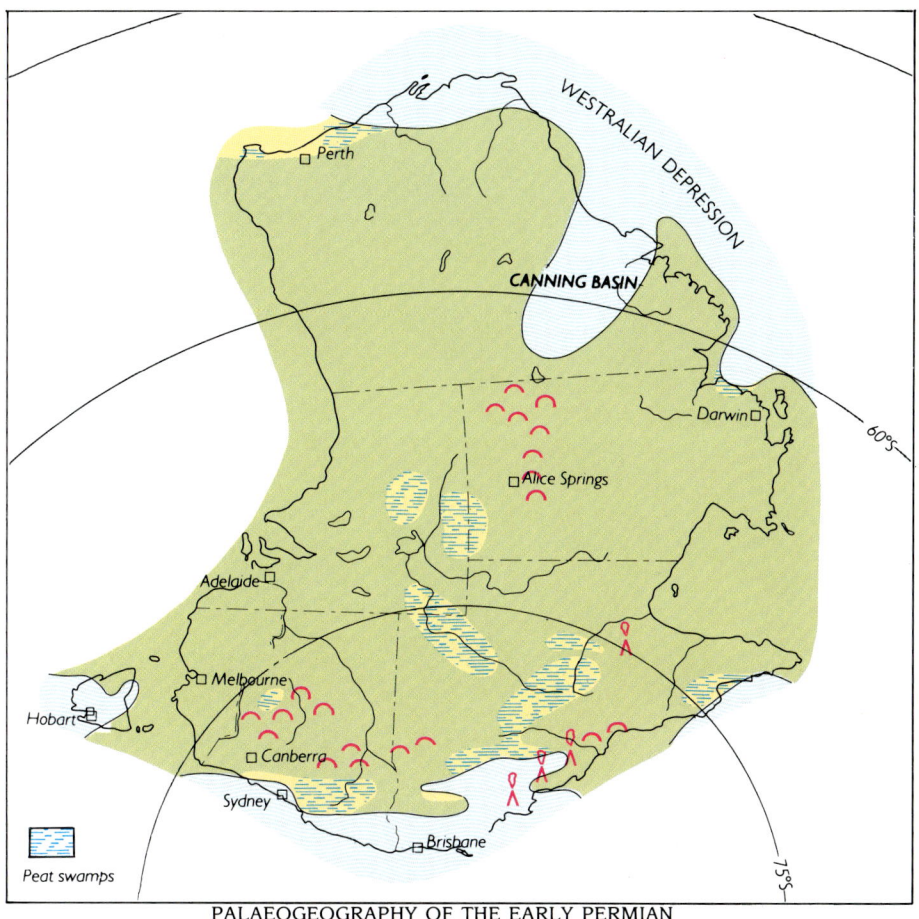

PALAEOGEOGRAPHY OF THE EARLY PERMIAN
270 million years ago

PALAEOGEOGRAPHY OF THE LATE PERMIAN
250 million years ago

"Lamp Shells", the commonest Brachiopod shell type. (Specimens MMF. Magn.X 0.5)

Permian Brachiopods from Yalwal, New South Wales. The Brachiopods were abundant throughout the Permian. They suffered many extinctions in the Permian Marine Collapse and were never again to attain the dominant status which they had held in the faunas of the Palaeozoic Era. (Specimen MMF. Magn.X 2.1)

better adapted Cycadophytes, Southern Conifers and Seed-ferns replacing the ancestral Glossopterids. The virtual disappearance of Glossopterids at the end of the Permian was in part due to this evolution into new plant groups. As well, the remaining stock was increasingly affected by contraction of their habitats. The cool-temperate coal swamps almost disappeared by the end of the Period. Thus it was environmental change which was the driving force behind the evolution of new forms better suited to the changing circumstances. Glossopterids had originated on the edges of an ice sheet in a cold world emerging from an ice age, flourished while conditions were within the range for which they had been designed, and in their turn gave rise to descendants better suited to the new conditions.

Glossopterids are the diagnostic plant fossils for the Permian in southern lands, and *Glossopteris* leaves are very common as fossils because of the deciduous nature of the plants which bore them.

The flora to which *Glossopteris* gives its name was rich and diverse. Ferns, Tree-ferns, Clubmosses, Horsetails, Ginkgophytes and Cordaites were abundant. The Cordaites had evolved into Conifers (excluding Southern Conifers) by late in the Period and disappeared from the Fossil Record in a manner similar to that of the Glossopterids.

Late in the Permian the hillsides and drier places were vegetated by Conifers, Cycadophytes and Ginkgophytes, all of which were no longer tied to watery environments. From this time onwards it seems likely that there were plants suited to most niches, as is the case today. It had taken about 150 million years from the time the first land-plants ventured onto the fringes of the land to reach this stage of a general vegetation cover for most of the ground.

A period of hot and humid conditions marks the Permian to Triassic transition. This abrupt climatic change altered the character of the vegetation. Suddenly Araucarian and Podocarp Conifers became abundant, small woody Mangrove-like Clubmosses grew in the deltas and estuaries, and the first Forked-frond Seed-ferns appeared.

The Permian marine Fossil Record reflects the changing environment of the times. Early in the Period faunas were impoverished high latitude biotas. On rocky sea bottoms along coastlines, *Eurydesma* bivalves grew in great abundance. Their strong shells fitted them for the high energy environments where wave action was violent. Their design was such that when they were knocked flat by waves they came up again, like childrens' toys which have a weight in the base and cannot be knocked over.

As conditions improved later in the Period, with warmer seas and an abundance of shallow-water environments due to the global sea level rise, there was rapid evolution of a diverse and rich fauna. By the middle of the Permian there were very large numbers of species, but at the end of the Period only about 10 per cent of them would survive and continue into the Triassic.

This Permian Marine Collapse appears to have been, in part, a reflection of the loss of shallow-water environments. It is interesting to speculate whether the coming together of Pangaea at this time might have contributed to the environmental changes which saw so many extinctions over a short time. The plate movements, with lands colliding and active tectonism, might have resulted in a concentrated episode of volcanism and climatic change, with acid rain and changes in acidity of sea water, in a manner similar to that which is postulated for the Terminal Cretaceous Event. The land vegetation of the time showed progressive changes and no cataclysm.

It has been found that the plant record is out of step with major faunal extinction events in the Fossil Record. Plants had, through the times, progressively adapted to the environmental changes which finally led to the animal extinctions. Their superior ability to hybridise and to produce polyploids and diploids meant that plant populations had variants available for selection when conditions changed. At the time of the Permian Marine Collapse the new flora of Southern Conifers, Cycadophytes and Mangrove-like Clubmosses which had already been established, made the transition

Glossopteris browniana from Blackmans Swamp, New South Wales. The prominent midrib and the lateral veins which form a mesh are characteristics of Glossopteris leaves. (Specimen MMF. Magn.X 1.3)

Tribranchyocrinus, the globular plated calyxes of Crinoids from the Permian Gerringong Volcanic Series in Kangaroo Valley, New South Wales. (Specimen AMF. Magn.X 1.6)

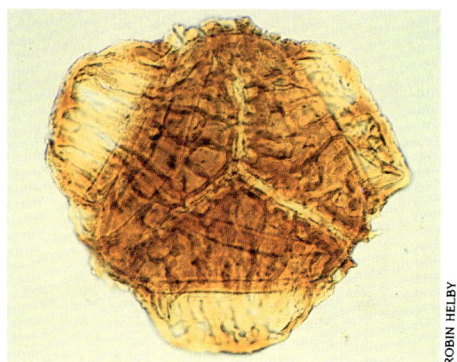

Dulhuntyispora maewestus, a microspore.

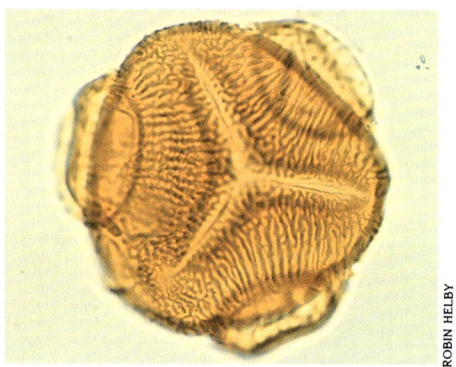

Dulhuntyispora dulhuntyi, a microspore.

Protohaploxypinus limpidus, a pollen grain.

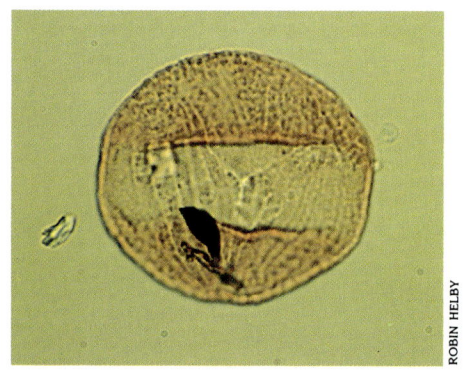

Marsupipollenites striatus, a pollen grain.

into the Triassic without visible dramatic changes. (Similarly, at the Terminal Cretaceous Extinction Event, the vegetation was changing rapidly to the modern-aspect type well before the Dinosaurs and other fauna disappeared. Consequently the plants show no interruption in their normal progression.)

The Permian Period was the time when Amphibians were still the dominant land Vertebrates, but Reptiles were abundant and fast overtaking them. The Vertebrate Fossil Record in Australia is meagre compared with that of other parts of Gondwana. In New South Wales, Labyrinthodont Amphibians which looked rather like giant Salamanders lived in the freshwater lakes and coal swamps of the Sydney Basin, preying on Fish. Reptilian footprints are known from the Late Permian Coalcliff Sandstone, but reptilian bones have yet to be discovered. (The first Mammal-like Reptiles were present in other parts of the world at this time.)

Insects were abundant worldwide, and Australia has several localities with rich and diverse Insect faunas. One such area is the Insect Beds at Belmont in New South Wales, where special conditions existed that led to perfect preservation of delicate organisms. There, volcanic ash had settled on the surface of a quiet lake, carrying down and entombing flying and floating Insects. The resulting rock formed from the compacted ash and mud on the lake bottom is so fine-grained that minute detail of the organisms fossilised in it is faithfully preserved.

NEW ZEALAND IN THE PERMIAN PERIOD

Permian rocks are widely distributed in New Zealand's South Island and also occur in the Northland region of North Island, where they are the oldest known rocks.

In some places the amazingly deep sedimentary marine series laid down during the Period is up to 20,000 metres thick and is one of the most complete Permian sequences preserved anywhere in the world. The trough in which the sediments accumulated on the outer edge of Gondwana was sinking rapidly during deposition. Folding of the strata was accompanied by intense volcanism and intrusion of igneous rock along the margins of the trough. Some of the basalt sheets were many thousands of metres thick. The sediments eroded from the Antarctica–Australia parts of the supercontinent adjacent to the trough were interleaved with basalt sheets and layers of sedimentary rock containing other volcanic material. The combined accumulation was subjected to deep burial.

Faunal successions of the Early Permian are similar to those of Australia but lack *Eurydesma* and some of the other species characteristic of this time in Australia. The presence of Corals of this age in New Zealand, and their absence in Australian waters, suggests that the waters where the trough was situated received warmer currents.

Some genera of marine Invertebrates in the New Zealand rocks have affinities with subtropical forms from the Tethys Sea.

Because of the alteration of Permian sediments by volcanism, deep burial, deformation and subsequent events, the somewhat sparse marine Fossil Record is hard to interpret. Rare *Glossopteris* leaves have been discovered in the extreme south of South Island, in sediments derived from the Antarctica sector of Gondwana.

There was no New Zealand land during Permian times. The country's Permian rocks which contain fossils were formed from sediments which collected in a trough on the outer edge of the supercontinent.

ROBIN HELBY

Permian Molluscs, *Myonia* (left front) and *Deltopecten* (middle right), from Maitland, New South Wales. (Specimen MMF. Magn.X 0.8)

Astralaster sp., a Permian Starfish, 260 million years old from Ravensfield, New South Wales. (Specimen AMF. Magn.X 2.0)

OVERLEAF:

A PERMIAN COAL SWAMP NEAR THE SOUTH POLE

About 260 million years ago

The *Glossopteris* Flora of the Permian swamps was inhabited by Labyrinthodont Amphibians and early Reptiles. Insects were also abundant. Glaciers still persisted on high mountains and the climate was cool-temperate with cold winters. A grove of *Glossopteris* trees is seen at the right of the picture, with Clubmosses, Ferns and Cycadophytes in the understorey. Cordaites are seen at left front, Conifers left centre, and in the right middle distance are Ginkgophytes and Tree-ferns. The Conifers, Ginkgophytes and Cycadophytes were able to live away from swamps and areas with high water-tables, and as a result much of the dry land was vegetated by this time.

A PERMIAN COAL SWAMP NEAR THE SOUTH POLE

PERMIAN FISH FROM BLACKWATER MINE, QUEENSLAND

Fish with a Glossopteris *leaf. (Specimens MQU, courtesy Michael Leu.)*

OPPOSITE PAGE:

Meganocrinus princeps, a very large Crinoid from the Branxton Formation in the Maitland district of New South Wales. The large hexagonal plates of the calyx are clearly seen, and the stem attachment is below. The arms are missing. (Specimen MMF. Magn.X 1.8)

Fenestella, a Permian Bryozoan. Bryozoans (Moss Animals) are small, colonial creatures which attach onto rocks, shells of other animals or Seaweeds. The units of the colony secrete tubes or boxes of lime, partially encasing their soft parts. Each organism looks like a Coral polyp, but is more highly evolved and organised. Bryozoans range from the Ordovician to the present day. (Specimen MMF. Magn.X 1.3)

An Early Permian Bryozoan from Allandale, New South Wales. (Specimen MMF. Magn.X 2.0)

Dicroidium callipteroides, the first of the forked frond Seed-ferns, which occurs with abundant Podocarp and Araucarian Conifers and with small Mangrove-like Clubmosses in the transition zone from the Permian to the Triassic. (Specimen AMF. Magn.X 0.9)

Footprints of a Labyrinthodont Amphibian. Large Amphibians were the dominant land Vertebrates of the Triassic. (Specimen AMF. Magn.X 1.2)

THE MESOZOIC (MEDIAN) ERA

The Mesozoic Era comprises the Triassic, Jurassic and Cretaceous Periods. Connecting the ancient and the modern Eras, it was the time when a vegetation dominated by Conifers, Cycads and Ferns was developed to its fullest potential and when the Reptiles ruled the land.

The first half of the Mesozoic saw the stabilisation of Pangaea, the "one Earth", in which the united landmasses of the world formed a supercontinent which stretched from Pole to Pole down one side of the globe. There was one mighty ocean, Panthalassa, which extended round some 70 per cent of the Earth at the Equator. As a result warm equatorial currents were free to flow, and travelled polewards along the shores. There were no circum-polar currents and thus no polar isolation and cooling, and consequently little gradation in temperature from Equator to Poles.

The Triassic Period was characterised by widespread aridity, more extensive and severe than at any other time in the Phanerozoic. A band of up to 50° on either side of the Equator was intensely arid. The Gondwanan lands lying in high latitudes were in a comparatively humid belt and the characteristic *Dicroidium* flora developed and thrived there. The environmental barrier posed by the arid low and middle latitudes prevented the mixing of southern and northern floras in the Triassic.

During the Jurassic and Early to Mid Cretaceous, world climates were uniformly humid and warm to hot. A great uniformity of vegetation developed all over Pangaea. Both Poles were luxuriantly vegetated. The climate at the Poles was warm-temperate and may even have reached subtropical at various times in the Cretaceous when world climates were hotter than at any other Phanerozoic stage. Globally, the vegetation comprised the ancient plant groups — Conifers, Cycads, Ferns, Ginkgophytes, Seed-ferns, Horsetails, Clubmosses and Mosses. Polar forests had to contend with several months of winter darkness even though the climate remained warm. Dinosaurs and other creatures that inhabited the forests were also adapted to living for long periods in the dark. Their food supply was constant, for the plants though dormant were still edible.

During the Cretaceous the break-up of supercontinents got under way and the continents began to move towards their modern positions on the globe. By late in this Period the fauna and flora started to change towards a modern-aspect biota in which Flowering Plants and Mammals prepared to assume dominance, and world climates started to be affected by the changing arrangements of land and sea.

Australia's geological history during the Mesozoic was a continuation of the evolution of a stable land. New Zealand's history as a landmass, however, only starts in the Mesozoic. Ancestral New Zealand was raised above the sea in the Late Jurassic. Up to that time its foundations had been laid down in marine basins and troughs on the edge of the Gondwanan supercontinent. Its formation and Mesozoic history are complex, and beyond the scope of this book. In summary, the pieces of crust which came together to form Ancestral New Zealand were "terranes", coming from several different sources and having different geological histories. During the Mesozoic the seafloor adjacent to the outer edge of Gondwana was not a simple ocean basin with one mid-ocean ridge, but was complex with several spreading ridges and several convergent margins. There may have been as many as five subduction systems with their axes at different angles. All the activities brought the terranes together, and they united and activity between them ceased. The amalgam thus produced was raised above the sea and the Ancestral New Zealand landmass came into being.

Cycas media growing near Darwin in the Northern Territory. The Mesozoic is commonly referred to as the Age Of Cycads because this group of plants was an important component of floras. Today Cycads are comparatively minor components of vegetation though locally abundant in some areas.

JIM FRAZIER

CHAPTER 6

LIFE IN THE WARM SOUTH-POLAR HEATHLANDS

THE TRIASSIC PERIOD
FROM 245 TO 208 MILLION YEARS AGO
DURATION: 37 MILLION YEARS

Much of the world was affected by aridity during the Triassic Period. Globally, temperatures were considerably higher than they are today, and temperature gradients from the Equator to the Poles were small. Due to its high latitude situation, Australia was insulated from the extremes of seasonal change and the climate was generally more humid. The South Pole was situated near present-day Bourke in western New South Wales for part of the Period.

The animals and plants of the Triassic were subjected to several months of darkness each year, and to ''midnight Sun'' for several more. They cast long shadows because of the low angle of the Sun, even in summer.

The Dicroidium Seed-ferns characterise the Triassic throughout Gondwana. They were the dominant plants in both heathlands and scrub on the fringes of the forest. South Polar Australia was well vegetated in the Triassic, with Conifers, Cycadophytes, Ginkgophytes and Tree-ferns in the forests, and Ferns, Clubmosses and Horsetails in the wetter habitats.

Labyrinthodont Amphibians dominated among land Vertebrates, but Reptiles were increasingly abundant. The first ancestral Mammals appeared in the Northern Hemisphere during the Triasssic.

There was no New Zealand land. Triassic rocks of New Zealand are marine, formed from sediments and volcanic material deposited on top of the deep Permian sequence in the trough on the outer edge of Gondwana.

Mesacridites elongatus from Beacon Hill in New South Wales. This long-winged Insect, a proto-Orthopteran, lived about 220 million years ago in the Triassic Period. (The Orthoptera are Grasshoppers, Locusts and Crickets, and this specimen is an early member of a group showing most of its diagnostic features. Magn.X 3.5)

Dubbolimulus, a Horseshoe Crab from the Triassic Ballimore red-beds in western New South Wales. (Specimen MMF. Magn.X 2.5)

The Triassic Period was characterised by heat and aridity over much of the globe. Evaporites occur in a band extending to 50° on either side of the Equator, and are more abundant and widespread than at any other time in the Earth's history. The middle third of the Period was the most intensely arid time, and most of the evaporite deposits were formed during this interval. The Early and Late Triassic were evidently as hot, but more humid. Red-beds were formed in profusion during these parts of the Period and are believed to be the result of alternating arid and humid conditions.

The distribution of reefs in Triassic seas parallels that of the Devonian and suggests that the temperature range was similar. Maxima were probably higher than today, and the seas were very warm.

The climate in high latitudes was less extreme. Warm-temperate and more humid conditions probably applied right to the Poles.

In the North America–Europe sector of Pangaea in the Late Triassic, the earliest and most primitive Mammals make their appearance in the Fossil Record. One such is *Megazostrodon rudnerae* which is a small rat-like creature only about 15 centimetres in length. Whole skeletons of this little animal have been found, showing its pointed snout and long tail. The Triassic Period was a long time before the Age Of Mammals, and the primitive species were to remain inconspicuous in the Reptile-dominated faunas of the Jurassic and the Cretaceous. Their time would come at the end of the Mesozoic Era.

AUSTRALIA IN THE TRIASSIC PERIOD

During the Triassic the South Pole was situated on the Australian sector of Gondwana. At this time the northern supercontinent Laurasia and the southern Gondwana were united into Pangaea.

Australia's high latitude situation insulated it from the extremely hot and arid conditions endured by many lands at the time. A warm-temperate climate, with long dry spells alternating with seasonal rainfall is the probable climatic pattern. Early Triassic floras, like that in the Narrabeen Group rocks in New South Wales, indicate that aridity was less of a climatic feature then than it was later to become. Narrabeen plants show less adaptation for drought resistance, having larger leaves and thinner cuticles.

Fresh water sedimentation was a feature of basins in the eastern half of the continent during the Triassic. In Queensland vast lake systems were gradually silted up with sediments derived from the river erosion of catchments. The lake and river systems of the Bowen Basin received sediments from the basin margins. In the Sydney Basin an enormous volume of sediment was carried by rivers from the New England highlands and then spread out over what must have been one of the last remaining floodplains like those of the Permian Period. These sediments formed the Narrabeen Group sedimentary rocks. The massive Hawkesbury Sandstone and strata formed in the Late Triassic were deposits comprising coarser sands, and had their origins not in New England but in the south. This derivation implies that there was high ground in the south-eastern sector of Australia or even in Antarctica. The coal swamps which had been such a feature of Permian landscapes had all but disappeared by Triassic times, with only small areas remaining at Leigh Creek in South Australia, in eastern Tasmania, and in north-eastern New South Wales and adjacent south-eastern Queensland. Volcanoes were active along the eastern continental margin, on the continental shelf, and on land east of the major basins.

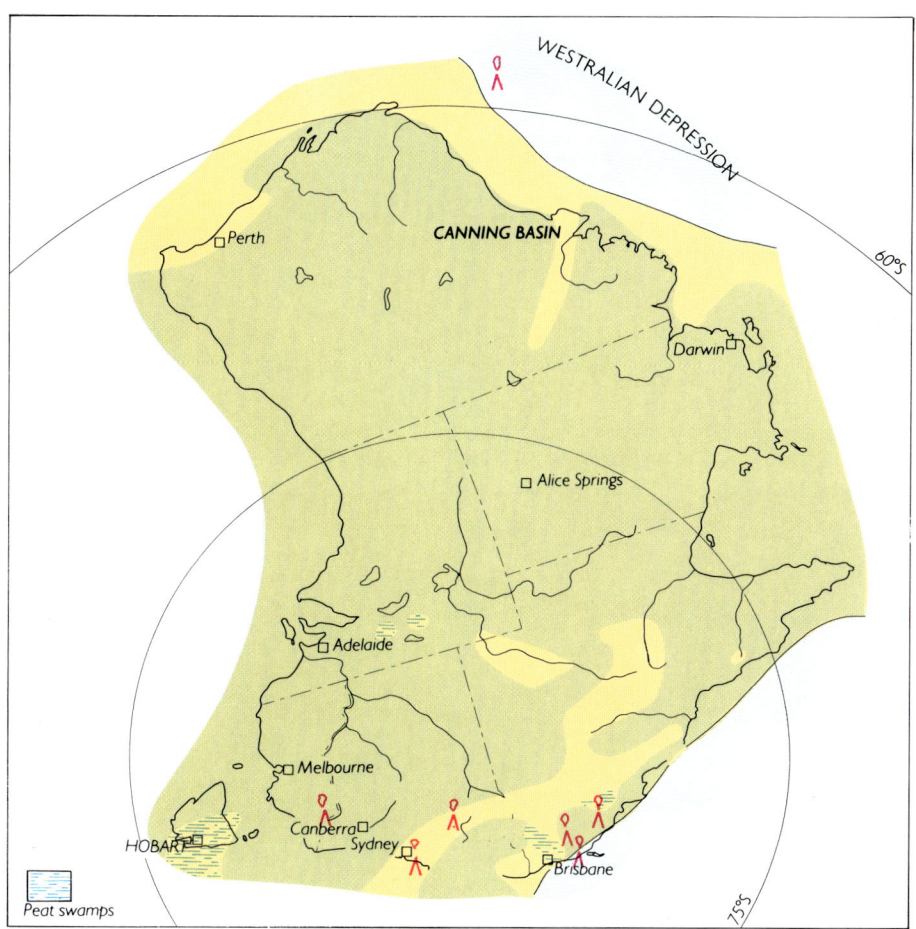

PALAEOGEOGRAPHY OF THE TRIASSIC
200 million years ago

In the Early Triassic there was a marine incursion down the Westralian Depression and onto the North West Shelf, and also on the Western Australian margin with two deeper penetrations inland. These penetrations were on the northern margin of the Canning Basin and into the Bonaparte Basin. The incursions had disappeared by the Mid Triassic, when the whole continent was emergent.

The North West Shelf marine transgression of Early Triassic times was followed by the establishment of a river system and the formation of deltas near the Exmouth Plateau, though the sea was to invade the region again late in the Period. The sandstones formed from sediments which were deposited during the Triassic constitute some of the main reservoirs for the natural gas fields in the area. Here, as elsewhere in Australia, it is interesting to note that after Devonian times the gas and oil deposits are largely terrestrial in origin, in contrast to those of other parts of the world which have mainly marine origins.

The Westralian Depression had become so choked by late in the Triassic that marine sediments only occur in its north-western sector. The repeated cycles of flooding and retreat by the sea, and the deposition of sediment in the depression, are typical of the events which occurred in the rifts along which separation between landmasses eventually took place.

The plants which characterise the Triassic in Gondwana are the *Dicroidium* forked frond Seed-ferns. Their fronds have not been found attached to trunks or larger stems, so their growth habit is unknown. It seems most likely that they formed a heathland vegetation on plains, while Conifers, Ginkgophytes, Cycadophytes and Tree-ferns formed forests and scrub on hillsides. Horsetails and small, woody Clubmosses grew on water margins.

Labyrinthodont Amphibians left their bones and walk-tracks in sediments containing the *Dicroidium* Flora. Amphibian-dominated faunas which contain rarer Reptiles have been found in many places in Australia. The

A Cockroach from Beacon Hill in New South Wales, from Triassic rocks about 220 million years old. (Specimen AMF. Magn.X 2.1)

TRIASSIC SPORES AND POLLEN

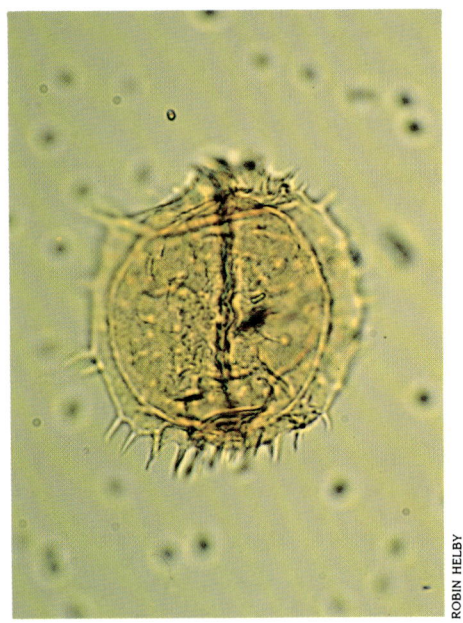

Aratrisporites tenuispinosus, a microspore.

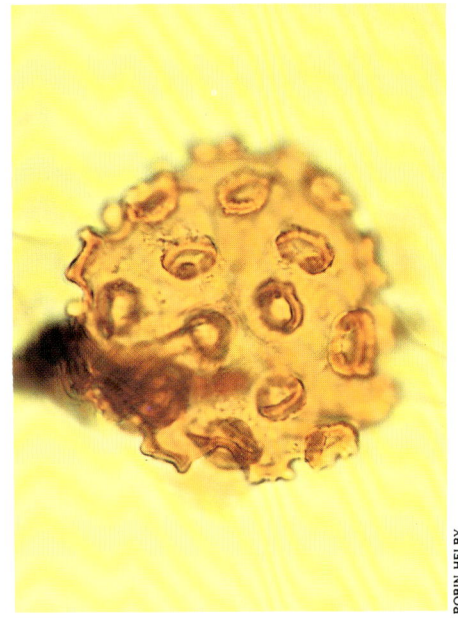

Craterisporites rotundus, a microspore.

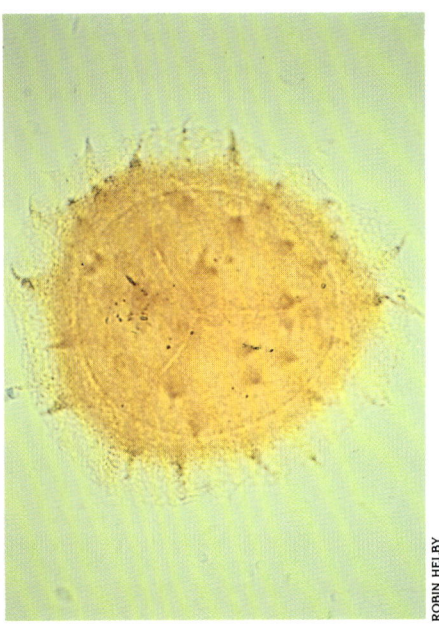

Kraeuselisporites cuspidus, a microspore.

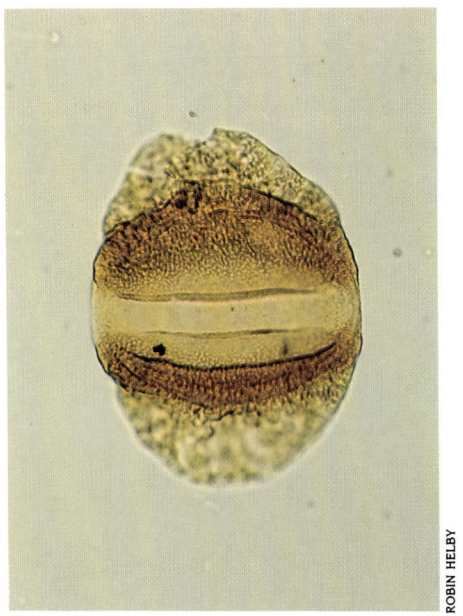

Falcisporites australis, a pollen grain.

Blina Shale in the Canning Basin of Western Australia has short-skulled Labyrinthodonts (like those that lived in the Sydney Basin in the Permian Period) as well as Capitosaurs which look like Crocodiles. Fragments of Thecodont Reptiles, a group ancestral to Dinosaurs, have also been found in the Blina Shale. They were either quadripedal, running on four legs, or bipedal, with two long back legs for rapid running in a semi-upright position and two smaller front legs. The Knocklofty Formation of Tasmania has a similar Amphibian fauna to that in the Blina Shale, as well as species of Thecodont Reptile closely related to South African and South-east Asian forms.

In Queensland the Bowen Basin's Arcadia Formation has a rich Amphibian fauna. A Thecodont and some Lizard-like Reptiles in the assemblage show affinities with the *Lystrosaurus* zone of southern Africa. (*Lystrosaurus* was a Mammal-like Reptile occurring in Africa, South America, Antarctica and India, but not found in Australia.) Footprints of Amphibians and Reptiles are also found in Triassic rocks in Queensland.

In the Sydney Basin in New South Wales footprints and bones of Amphibians are found in Triassic rocks. A spectacular skeleton of a Capitosaur *Paracyclotosaurus* was found in the St Peters brickpit in an inner-city suburb of Sydney.

Freshwater Fish were abundant in the lakes and rivers of the Early Triassic. The Sydney Basin has some remarkable fossil Fish localities. An Australian Museum excavation at Somersby, near Gosford in the late 1980s collected fossils of Lungfish, Sharks, a long-nosed predator *Saurichthys* and large numbers of the common *Cleithrolepis*.

At the famous Beacon Hill Quarry locality at Brookvale in suburban Sydney, a specimen of a Horseshoe Crab, *Austrolimulus*, has been found along with Insects, plants and bones of large Amphibians. (Another specimen of a Horseshoe Crab was collected in the late 1980s near Dubbo, in western New South Wales and named *Dubbolimulus*.)

Very few marine sedimentary rocks of Triassic age are found in Australia, so the marine Invertebrate record is poor. It does, however, contain Ammonites with cosmopolitan affinities, which are useful in determining the

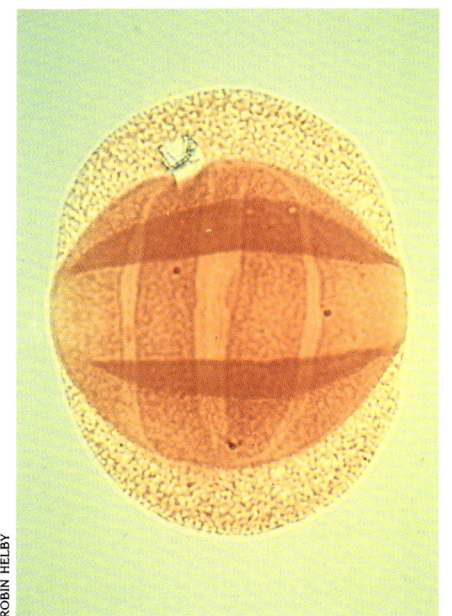

Lunatisporites noviaulensis, a pollen grain.

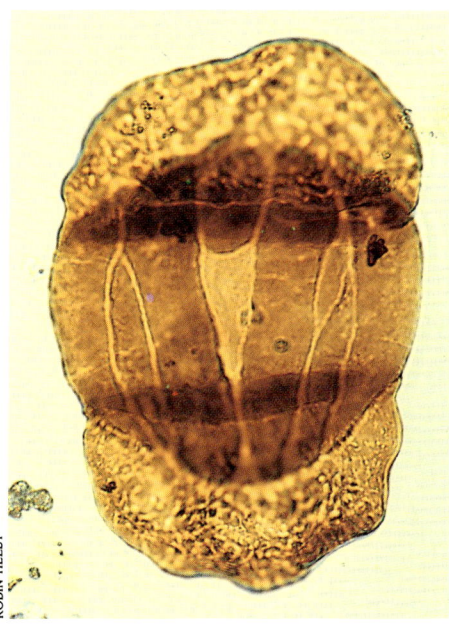

Protohaploxypinus samoilovichii, a pollen grain.

Monotis, a common Triassic Bivalve Mollusc.

age of different horizons. (Many species of Invertebrates achieved near-global distribution during this time of one landmass and one ocean, when seas were universally warm.)

Dinoflagellates — single-celled planktonic organisms which are classified within the Plant Kingdom because they contain chlorophyll, but whose other characteristics make them more like animals — first appeared in the Triassic, and they continue to be abundant in the waters of the globe today.

NEW ZEALAND IN THE TRIASSIC PERIOD

Rocks of Triassic age are widely distributed on both main islands of New Zealand. Apart from a few formations right at the top of the sequence, whose origins are uncertain, all these rocks were laid down under the sea. During the Period there may have been some ephemeral fragments of land in the marginal province that formed the outer edge of Gondwana. A volcanic arc on the outer margin of the deep and sinking trough, in which the vast depth of sediments had accumulated in Permian times, contributed ash and volcanic material to the Triassic sediments accumulating in the trough. The volcanoes were probably largely submarine, and their activity declined as the Period progressed.

Plants found in Mid Triassic rocks of Canterbury and northern Otago indicate that the Gondwanan shorelines were not far distant when the fossiliferous sediments were laid down. Deltas on the edge of the Gondwanan supercontinent probably provided the sediments. A typical Triassic flora with *Dicroidium*, other Seed-ferns, Ferns, Ginkgophytes and Conifers is present.

Marine faunas of the Triassic are rich in Ammonites, Bivalves and other sea creatures, many showing affinity with European species. (This time of one ocean and uniformly warm water conditions encouraged uniformity.) No terrestrial Vertebrates were preserved in the sediments, but a marine Reptile, *Ichthyosaurus*, has been found in the Mt Potts region.

OVERLEAF:

A TRIASSIC SCENE IN THE SYDNEY BASIN
About 225 million years ago

The Sydney Basin was part of a large floodplain with lagoons and deltas near the coast. There was seasonal aridity and much of the sandy country was sparsely vegetated. The swamps and water margins had a good cover of Horsetails, Seed-ferns and Ferns, and abundant woody Clubmosses which looked like dwarf Palms were the "Mangroves" of the times. Conifers, Ginkgophytes and Cycads were adapted to living on the dry hillsides. Dragonflies, Mayflies and other Insects were abundant (those in the picture are not drawn to scale). A Labyrinthodont Amphibian is seen at left, and an early Crocodilian Reptile at right.

133

A TRIASSIC SCENE IN THE SYDNEY BASIN

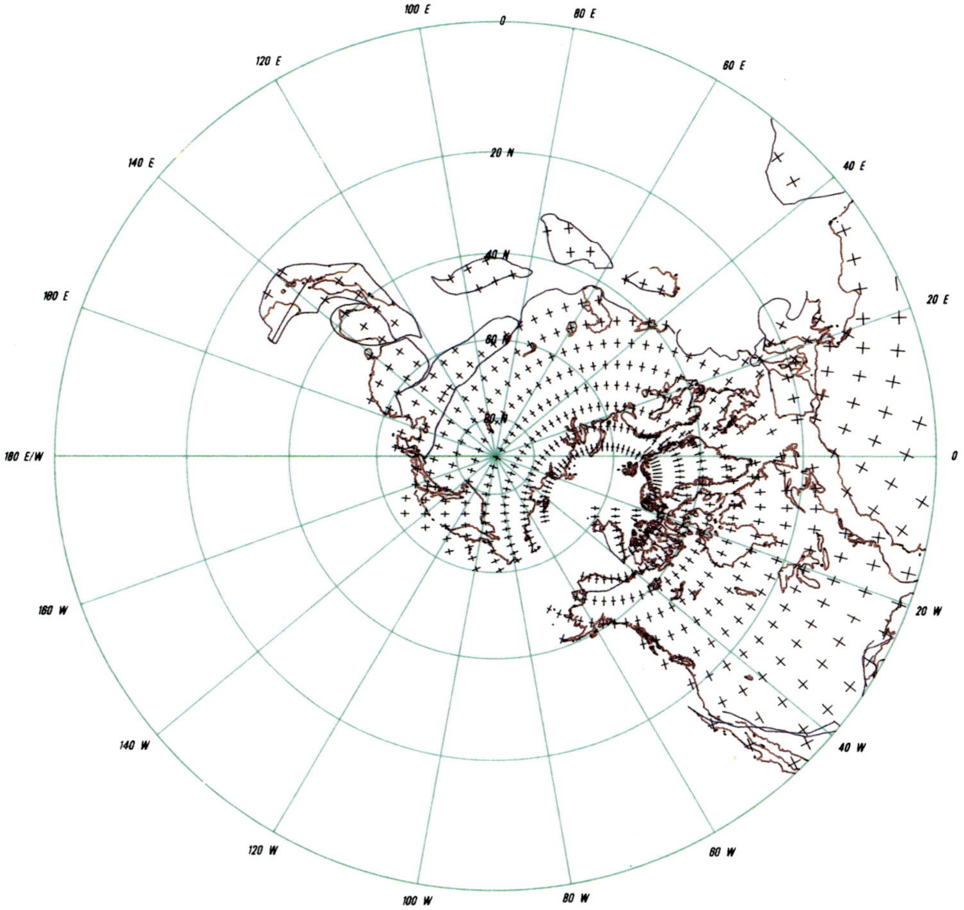

NORTH POLAR PROJECTION

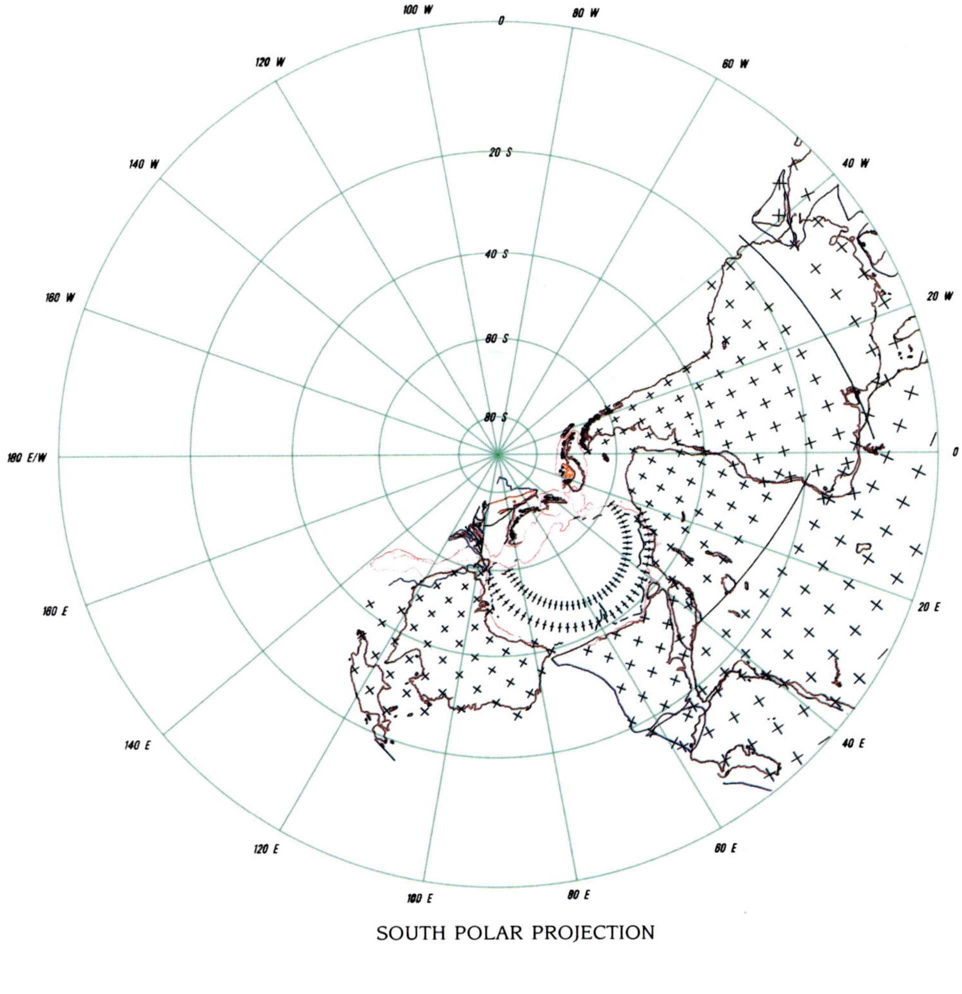

SOUTH POLAR PROJECTION

OPPOSITE PAGE:

TRIASSIC INSECTS FROM BEACON HILL, NEW SOUTH WALES

A Beetle. (Magn.X 2.5)

A winged Insect. (Magn.X 5.2)

Clatrotitan andersoni, a large insect wing. (Magn.X 1.7)

Dicroidium from Dinmore, Queensland. The forked frond Seed-ferns of the genus Dicroidium characterise the Triassic in lands which were once united in Gondwana. (Specimen AMF. Magn.X 1.0)

POSITION OF LANDMASSES IN RELATION TO THE NORTH AND SOUTH POLES IN THE TRIASSIC PERIOD, 220 million years ago

CHAPTER 7

A WARM WORLD WITH DINOSAUR FORESTS

THE JURASSIC PERIOD
FROM 208 TO 144 MILLION YEARS AGO
DURATION: 54 MILLION YEARS

While still in high latitudes, in a warm to hot world with generous rainfall, Australia shared with the rest of Pangaea a luxuriant vegetation dominated by Conifers, Cycads and Ferns. Preparations were under way for the fragmentation of the supercontinents. Rift valley systems formed along the margins of continents. Ancestral New Zealand started to rise above the sea in the middle of the Period, and the Rangitata Orogeny of the Late Jurassic-Early Cretaceous accelerated the production of high ground.

Dinosaurs dominated the Vertebrate faunas. It was a time of abundance for plants and animals.

Agathis jurassica and Pentoxylon australica from the Jurassic Talbragar Fish Beds in New South Wales. Small deciduous twigs of Agathis with a number of leaves, representing a season's growth, are common fossils at Talbragar. These ancient Kauri Pines dominated the forest 175 million years ago. Pentoxylon is a Cycadophyte with long, narrow leaves.

Pyramid Rock, on the southern tip of Maria Island, Tasmania. With the start of rifting between the lands which comprised Gondwana, there were outpourings of dolerite and basalt along the lines where the fractures were occurring. Tasmania was in a hinge position between Antarctica and Australia, and was almost covered by floods of igneous rock over time. Today many of its landscapes bear dramatic evidence of this phase of its geological history. In particular, black cliffs with prominent vertical joints stand out along the coastline.

At the start of the Jurassic Period the landmasses of the world were united into Pangaea. Most of the continental crust was above sea level, though part of western North America was covered by shallow seas. The Sierra Nevada mountains were rising on the margin of this submerged area during the Period.

Preparations for the fragmentation of supercontinents began during the Jurassic. The start of rifting was marked by outpourings of basalts and dolerites on rift lines. These Jurassic volcanic rocks are a feature of all southern lands and often survive today as landscape features — notably in Tasmania at the mouth of the Derwent River.

From the Mid Jurassic onwards, the processes which established the present-day plates, continents and oceans were well under way. The outline of Australia, the island-continent, was to be delineated during this time. The eastern margin of Gondwana, comprising the Ancestral New Zealand landmass, was raised above the sea by the end of the Period.

Climates worldwide were warm to hot and there was no polar ice. High latitude climates were warm and humid. More seasonal climates in which long dry spells alternated with rainy periods characterised a belt which extended possibly to 40° on either side of the Equator.

On the world scene, the Jurassic was the Age of Dinosaurs. For the Plant Kingdom, it was the Age Of Conifers And Ferns. Conifers, Cycads, Ginkgophytes and Seed-ferns (all of which are gymnospermous plants), together with ground Ferns, Tree-ferns and the Fern-allies, the herbaceous Horsetails and Clubmosses, comprised the vegetation. A similar vegetation composed exclusively of these ancient plant groups was to persist into the Early Cretaceous. There is some evidence from a study of Laurasian floras that the diversity of plants was declining towards the end of the Jurassic. This decline may suggest that the vegetation had reached a stage of stagnation. Under the very long period of unchanging conditions which had applied, the genetic ability to change was gradually being eroded. Populations which reach this stage are headed for challenges to their dominance when conditions change substantially and new and efficient forms arise to meet the new conditions. This stagnation may have contributed to the rapid rise to dominance of the Flowering Plants in the Late Cretaceous.

In the marine Microfossil Record Dinoflagellates and Foraminifera were abundant in plankton. Among Invertebrates the Ammonites underwent great development, with modern Ammonite families becoming established and some of the primitive families disappearing. Belemnites became abundant. On the land Insects continued to abound and diversify, with Flies and Bees making their appearance in the Fossil Record.

AUSTRALIA IN THE JURASSIC PERIOD

Australia still lay in the high latitude, humid belt during the Jurassic Period. Long months of winter darkness were the only adverse factor to which the fauna and flora had to adapt.

The Jurassic was an important time for the formation of petroleum deposits in Australia. Source rocks were laid down in the Barrow Island oilfield in the Carnarvon Basin of Western Australia, and source and reservoir rocks in a number of other fields. These others include the natural gas and condensate fields in the Surat Basin of Queensland, the Jackson Oilfield and smaller fields in the Eromanga Basin of South Australia and

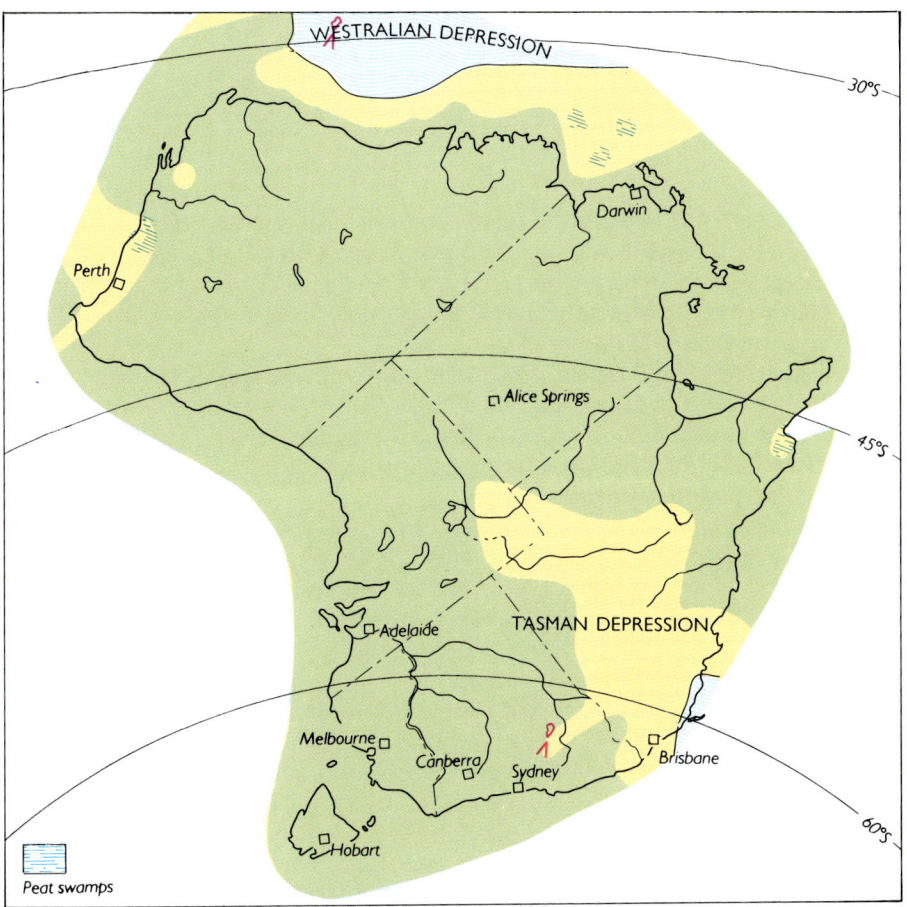

PALAEOGEOGRAPHY OF THE EARLY JURASSIC
200 million years ago

Agathis macrophylla, a living species of Kauri Pine.

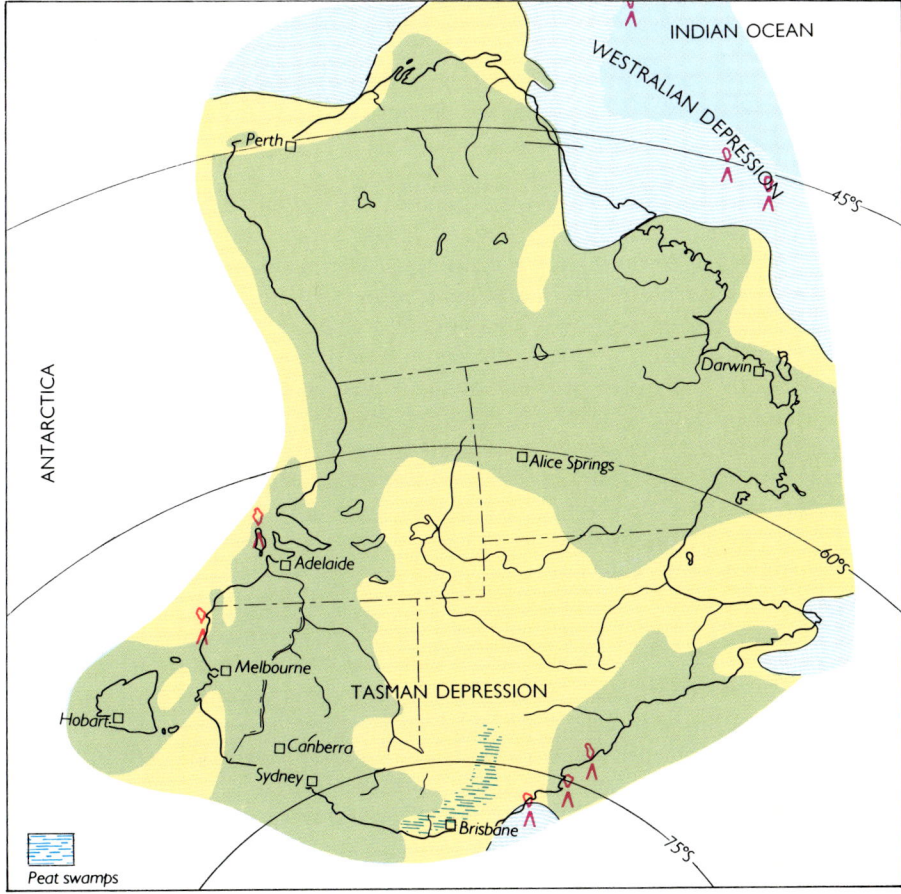

PALAEOGEOGRAPHY OF THE MID JURASSIC TO EARLY CRETACEOUS
170 to 140 million years ago

Leaves of the type belonging to Bennettitalean Cycadophytes which have polycarp fruiting structures like that from Lune River. (Specimen BMR. Magn.X 1.0)

These two sections were taken from the same Pentoxylon trunk at a distance of about 30 centimetres apart. They show the rapid tapering of the trunk. It is unknown if the broad basal part with its more complex structure represents a lignotuber or if the trunk was simply buttressed or widened out into a bulbous base above ground. (Specimens from Chinchilla, Queensland, kindly loaned by John Bennett of Perth. Magn. X 1.2)

adjacent States, and sites in the Gippsland Basin of Victoria. A group of planktonic micro-organisms, the Dinoflagellates, became abundant in the ocean basins which were forming as rifting proceeded around the margins of the continent. The Dinoflagellates contributed to the organic matter which later produced hydrocarbons.

The first evidence of impending crustal movements heralding the start of rifting to free the Australian landmass from Gondwana is seen in the dolerites and basalts which erupted along rift lines. These volcanic rocks, being resistant to erosion, are dramatic features of some landscapes. They are particularly evident in Tasmania.

Preparations for Australia's separation from the supercontinent started when crustal fragments flanking the Western Australian coast started to move away. It is possible that western Thailand once lay beside the north-western sector of the Westralian Depression where the seafloor spreading started, however there is no conclusive evidence to prove the identity of this neighbouring fragment.

A marine transgression occurred over the north-western margin of the continent late in the Period. There had been rifting and faulting for some millions of years preceding the seafloor spreading. The rifts did not occur along a single line, making a neat fracture, but involved a tearing and shearing component resulting in parallel ridges and depressions at an angle to the continental margin. These basin structures, later infilled with sediment, acted as traps for hydrocarbons. The Rankin deposits in the Westralian Depression are examples formed in this way. Activity on the major Darling Fault in the Perth Basin resulted in sediments accumulating in a deep trough, and there was a brief marine incursion into this trough in the Mid Jurassic.

Similar faulting and rifting along Australia's southern margin, from the Mid Jurassic, was setting the stage for separation from Antarctica, though this link was to be the last to be severed at the completion of the Gondwanan break-up.

Volcanic vents, or diatremes, were active near the eastern margin of the Sydney Basin late in the Jurassic. This volcanic activity was probably connected with the impending opening up of the Tasman Sea which separated the outer margin of the supercontinent from the main landmass.

The Tasman Orogen became effectively stabilised during the Jurassic and was undergoing erosion by major river systems flowing generally eastwards and entering the sea north of Brisbane. This vast tract of country was a broad series of interconnected depressions, each with its own river, lake and swamp system. The conversion of all these systems to the major structural unit, the Tasman Depression, took place between the early part of the Mid Jurassic and the earliest Cretaceous times in five tectonic cycles associated with the disintegration of eastern Gondwana. As the depression subsided, high ground was raised to its north-west and to its south in five pulse-like episodes. In the four of these cycles encompassed by Jurassic time, catchments were elevated and the rejuvenated rivers eroded the high ground and contributed coarse sediments to the depression. As the rivers became less active with the lowering of the catchments and the silting up of the lakes and streams and swamps, erosion slowed, allowing fine sediments to accumulate. Then another uplift caused a further rapid phase of erosion. In the fourth cycle, at the end of the Jurassic, lagoons and deltas formed along the north-eastern margin of the depression. They developed as a result of a global sea level rise which affected drainage patterns. (The fifth cycle in earliest Cretaceous times was to be cut short by a rapid rise in sea level, which caused flooding of the whole depression.)

A volcanic arc lay to the east of the Gondwanan supercontinent on a

A Jurassic Sponge. (Specimen AMF. Magn.X 1.8)

A Sand Dollar Echinoderm from South Australia, of Jurassic age. (Specimen U.NSW. Magn.X 1.4)

Leptolepis talbragarensis, a small freshwater Fish common in the Talbragar Fish Beds. (Specimen MMF. Magn.X 1.8)

zone of interaction between the proto-Pacific and Gondwanan plates. The outer margin of Gondwana was bent into an S-shape (orocline) in the Tasmania-Antarctica region during the tectonic events which from Mid Jurassic times started to raise the Ancestral New Zealand landmass above the sea. Volcanic ash from the arc contributed to sedimentation in the Tasman Depression.

The sedimentary sequences formed in the Tasman Depression included the main aquifers (water holding strata) for the Great Artesian Basin, and reservoir rocks for the accumulation of oil and natural gas in the Roma, Moonie and Jackson fields. The alternating types of sediment resulting from cycles of rapid erosion followed by quiet deposition during the phases of tectonic activity in the Tasman Depression produced the two types of sedimentary rock required for the formation of hydrocarbon deposits: porous source and reservoir rocks of coarse-grained or unsorted sediment to hold the deposits; and cap rocks from fine muds or silt to seal them.

Coal swamps developed along the lower reaches of eastward-flowing rivers in north-eastern New South Wales and south-eastern Queensland in the Mid to Late Jurassic. There were also coal swamps north of Perth in Western Australia at this time.

The environment of the Tasman Depression in Jurassic and Early Cretaceous times has no counterpart in Australia today. The Amazon Basin is its closest modern analogue. It was a vast area of low relief with large sluggish rivers, and lakes and swamps. Under the warm and humid conditions of the times, the region was luxuriantly vegetated. The composition of the vegetation was very different from that of the Amazon now. There were no Flowering Plants to form the Broad-leaf Rainforest. Instead, the forest vegetation comprised Conifers, Cycads, Ginkgophytes and Ferns. Long dark winters, summers without night, and the low angle of the Sun made the Tasman Depression in the Jurassic unique.

The vegetation in Jurassic times in Australia comprised an abundance of Southern Conifers including Kauri Pines (Agathis), Araucarians (Monkeypuzzles similar to Hoop, Bunya and Norfolk Pines) and Podocarps similar to those alive today. Pentoxylon Cycadophytes — which may be ancestral to the living Screw Pine, Pandanus (a monocotyledonous Flowering Plant) — are common as fossils. There were also plants of a group of Seed-ferns which may have been ancestral to other Flowering Plants.

Jurassic plant fossils occur at many localities in Australia, and some sites are famous for the beauty of their specimens. One notable site is the Talbragar Fish Bed locality, in western New South Wales, where a rich flora occurs in association with freshwater Fish and a single example of a Cicada. It is possible to reconstruct the Mid Jurassic scene of 175 million years ago when the Talbragar Fish were alive. The sediments in which the fossils are preserved were accumulated in a freshwater lake of limited area. This small lake lay in a forest of Kauri Pines (Agathis jurassica) in which some Podocarp Conifers were dispersed. Pentoxylon Cycadophytes formed an understorey. A heath zone of Seed-ferns (Pachypteris) surrounded the lake. Tree-ferns and ground Ferns grew in suitable places. Forests like this ancient Talbragar example still exist today. Indeed, the Atherton Tableland in Queensland has Rainforest remnants with dominant Agathis (Kauri) intermingled with Podocarp Conifers. A Cycad (Lepidozamia hopei) grows in the understorey as a Palm-like plant attaining considerable height. Tree-ferns and ground Ferns are also present.

Another famous locality is that at Lune River in Tasmania, where a basalt flow overlies a zone in which abundant petrifactions occur. These petrifactions are agatised Tree-fern trunks, tree trunks of Conifers, and rare

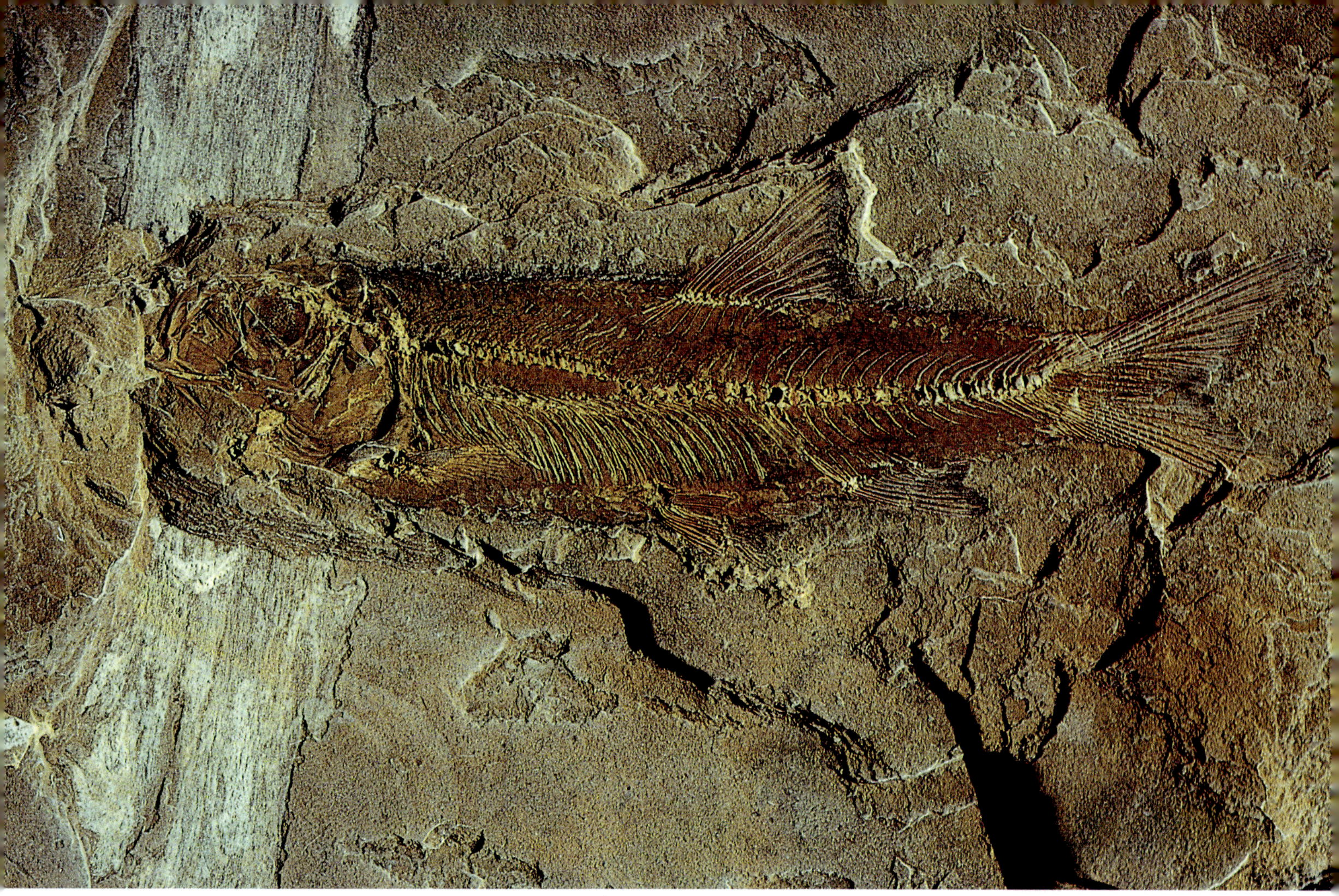

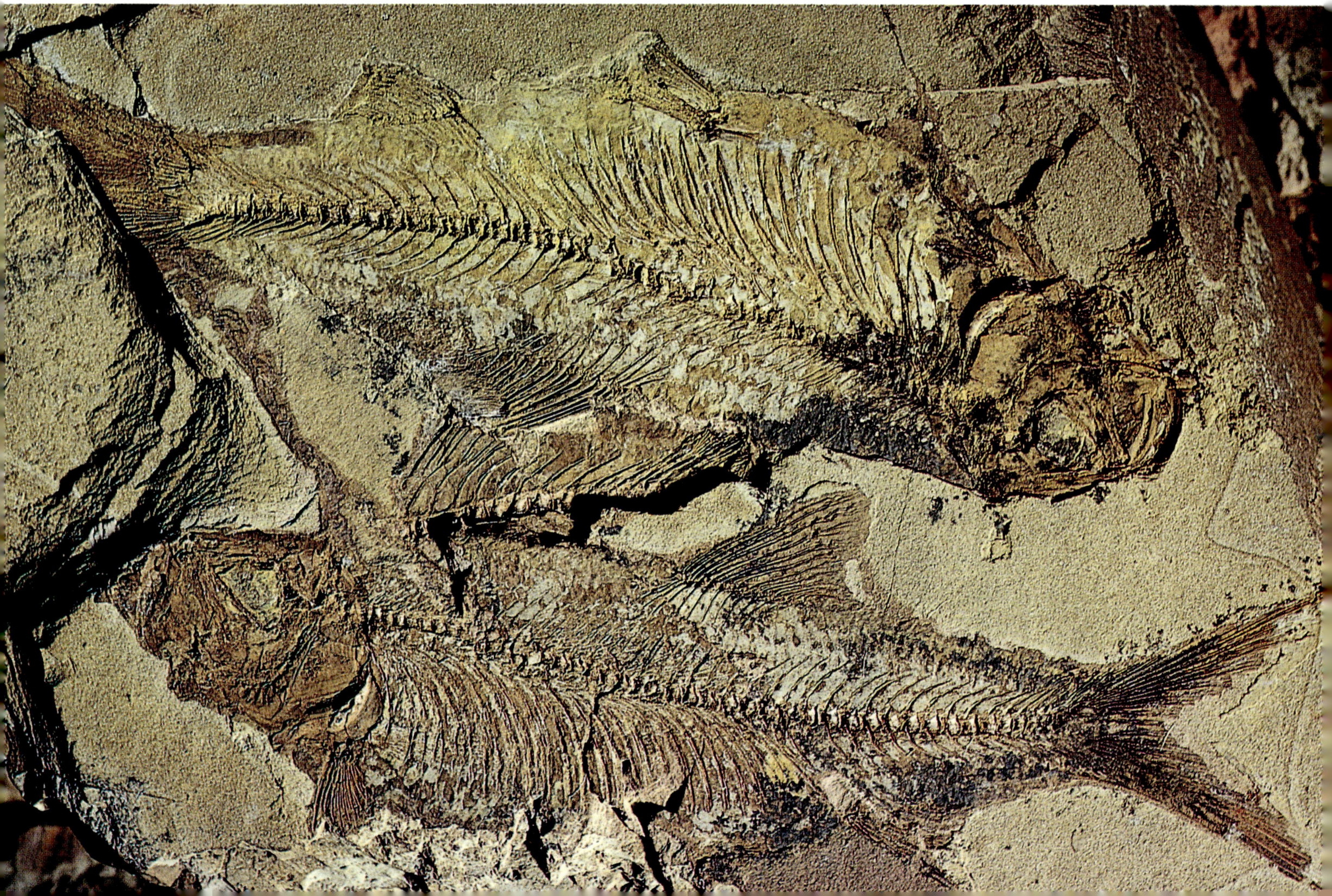

examples of Bennettitalean Cycadophyte fruiting bodies in which molecular replacement of plant material with mineral matter results in the complete preservation of cellular structure. Many beautiful agates, including picture agates, also occur at this site. The beauty of all the specimens — fossil and mineral — found at Lune River has attracted so many collectors that unfortunately little remains *in situ*.

Despite the months of winter darkness, Dinosaurs lived in Australia during the Jurassic. There would have been no shortage of food as the plants on which they depended were not deciduous and would have been merely dormant during the dark months when photosynthesis ceased. The enormous Sauropod *Rhoetosaurus* has been found in freshwater deposits in Queensland, and is one of the oldest Sauropods known. Queensland deposits also contain Theropod Dinosaurs of different sizes, Dinosaur footprints, and remains of Plesiosaurs (large aquatic Reptiles).

Australia has the only known Jurassic Labyrinthodont Amphibian; elsewhere in the world Labyrinthodonts disappear from the Fossil Record in the Late Triassic. The Queensland Kolane Amphibian, *Siderops kehli*, was a large creature with a body about 2.5 metres long and a broad, flattened head about 0.5 metre long and wider than it was long. It had a mouth full of fierce teeth and was probably an aggressive hunter.

The Jurassic marine Invertebrate record is scanty. Sediments laid down in the Westralian Depression during marine incursions contain a shallow-water fauna with Ammonites, Belemnites, Bivalves and a rich microplanktonic assemblage.

Living Dinoflagellates photographed by a scanning electron microscope show the complete form of the organisms. Fossils of Dinoflagellates are only of the cysts. This living species causes Shellfish poisoning in humans.

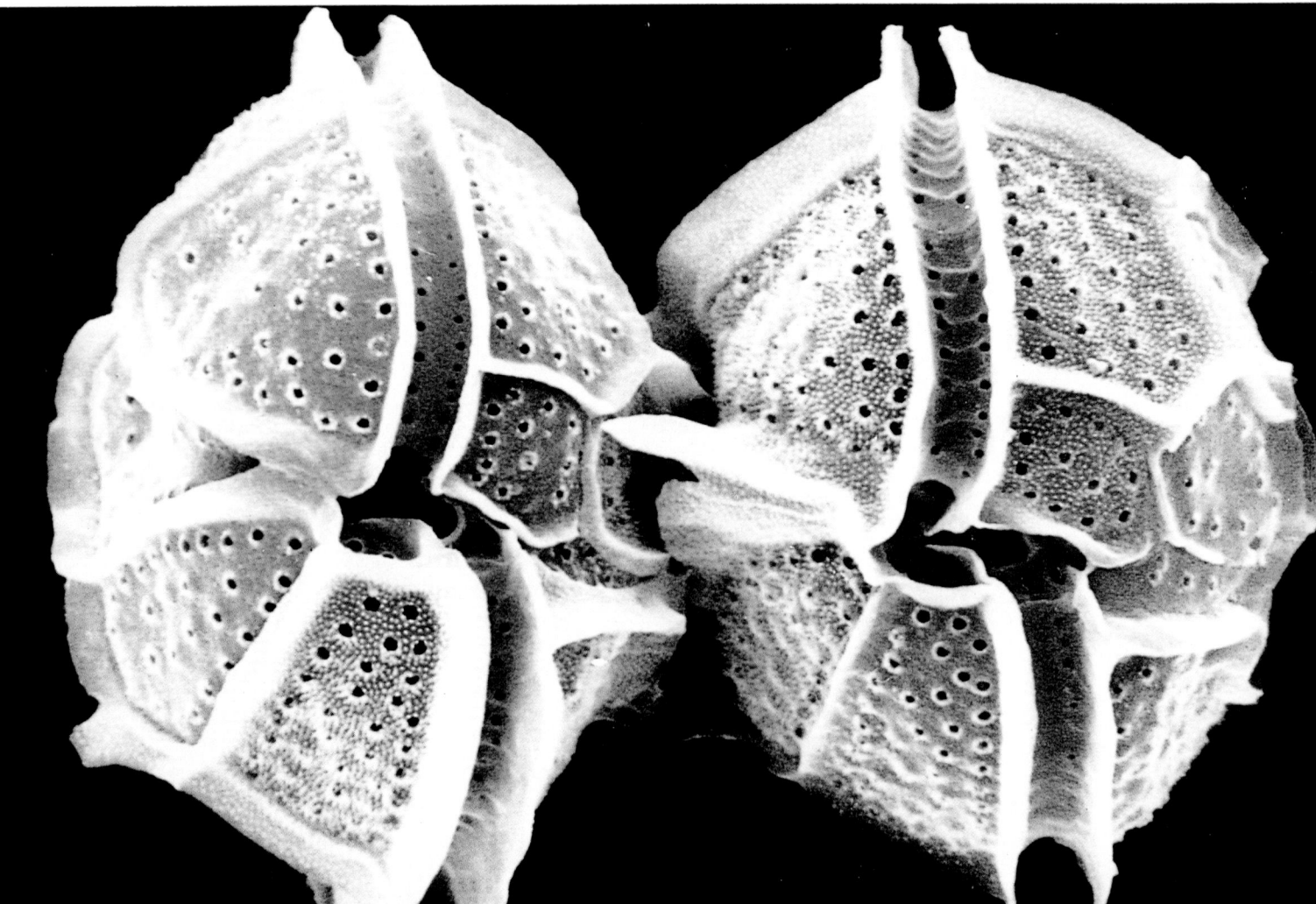

NEW ZEALAND IN THE JURASSIC PERIOD

The Jurassic rocks of New Zealand are mainly the result of the continuing deposition of sediments on top of those of the Late Triassic, which had accumulated in the deep trough offshore from the margin of Gondwana. The sediments were predominantly sands, muds and conglomerates. From the Mid Jurassic, tectonic movements started to raise a large landmass above the sea. This tectonic episode, which was accompanied by volcanic activity, caused the collapse of the trough system. In the Late Jurassic non-marine and estuarine deposits make their appearance.

Ancestral New Zealand straddled the zone where interaction between the proto-Pacific Plate and the Gondwanan margin was occurring. Australia's slight clockwise rotation round Antarctica during early preparation for its separation had created a bend (which was roughly S-shaped, and known as an orocline) at the edge of the plate. The orocline predetermined the shape of modern New Zealand, which still lies on the curved line where plates meet. As the zones where plates meet are the earthquake and volcano regions of the world, the islands of New Zealand are decidedly "the shaky isles".

By the latest Jurassic times the emerging landmass had attained considerable size. It occupied an area from the Campbell Plateau to New Caledonia in length, and from the Chatham Islands to the Lord Howe Rise in width. The absence of marine rocks of Late Jurassic and Early Cretaceous age in New Zealand today is evidence that by then its whole area was dry land.

The Early Jurassic flora in New Zealand includes Podocarp Conifers, Horsetails, Cycadophytes and Ginkgophytes. *Cladophlebis* Ferns and *Pentoxylon* leaves are abundant from the Mid to Late Jurassic. The flora is the same as that found in Australian and Antarctic Jurassic strata.

The accumulation of Mid Jurassic tree trunks found fossilised at Curio Bay in Southland suggests that warm-temperate conditions compatible with luxuriant forest growth prevailed at the time. Such climate is known to have then existed in adjacent Gondwana.

The oldest coal mined in New Zealand is found in Southland. It was formed from the carbonaceous material produced by a rich vegetation which had become established on the emerging land by the Mid Jurassic.

New Zealand has a relatively rich and diverse Early Jurassic marine fauna of Molluscs, Brachiopods and other groups (in contrast to Australia where marine beds of this age are poorly represented). The marine forms are of an isolated Maorian provincial type, not closely related to those of the rest of the Pacific region. Marine fauna from the middle of the Period shows an affinity with that of Australia, but by the Late Jurassic provincialism is again evident. The massive tectonic movements which were occurring no doubt influenced currents and created locally isolated and special conditions.

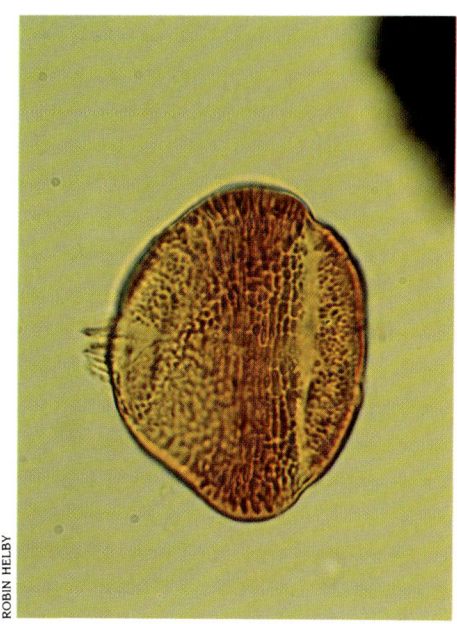

ROBIN HELBY

Corollina torosa, a pollen grain.

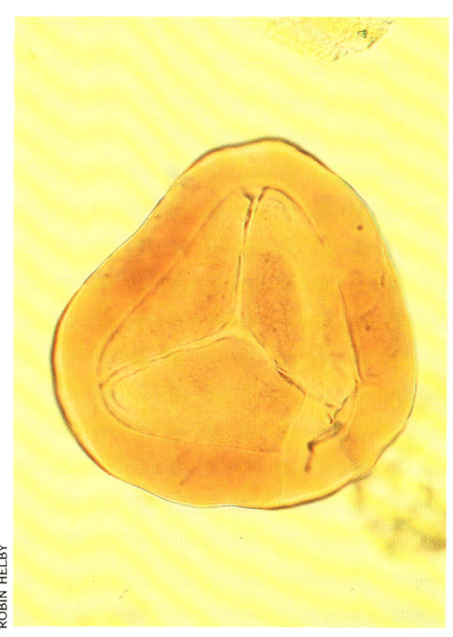

ROBIN HELBY

Murospora florida, a microspore.

DINOFLAGELLATES

Dinoflagellates are microscopic, mobile, flagellate, acellular organisms which are classified in the Plant Kingdom in an algal group, the Pyrrhophycophyta, because they have a photosynthetic pigment. They have a fossil record because they form cysts composed of a complex protein substance called sporopollinin, which is the material also found in the exines of pollen and spores. This material is resistant to decay, as evidenced by the plant Microfossil Record which goes back to the earliest land-plants more than 400 million years old. Dinoflagellates only enter the Fossil Record in the Mesozoic.

Dinoflagellates, in their active state, have a theca (outer cover) made of cellulose, comprising a number of plates, inside which the living protoplasm is contained. Two projections form flagellae, which are whip-like threads that emerge through holes in the theca, one lying in a transverse groove and the other trailing. The flagellae propel the organism through the water with beating movements. The cell contents inside the theca contain the nucleus and the chromoplasts (which are brownish or yellowish organelles containing chlorophyll for photosynthesis). Some Dinoflagellates have light-sensitive eye-spots to keep them in the photic layers of the sea.

When the Dinoflagellate is no longer active, because of a change in temperature or other factors, it becomes enclosed in a cyst. The theca dissolves and the living material contracts towards the centre of the cyst, connected to the cyst wall by radiating projections. The cyst wall of sporopollinin is usually also made of plates, corresponding to the plates of the theca which it replaced. When conditions improve, the cyst loses a plate to allow the protoplasm to emerge and form another theca. Some Dinoflagellates are not photosynthesisers, having lost the pigmented organelles. They are saprophytic or ingest other organisms, behaving like members of the Animal Kingdom. Among unicellular Algae and acellular (Protistan) animals are many "intermediate" organisms whose classification is somewhat arbitrary.

It has been recently discovered that some Dinoflagellates practise chemosynthesis when living in the dark, and can thus produce food without relying on photosynthesis. This ability is most interesting when one considers the vital role of phytoplankton in the marine food chain. It has to be wondered if some "intermediate" planktonic organisms forming the basis of food chains in high latitudes, where there is darkness for several months of the year, might have the ability to make food in the dark. This ability might be more widespread than has been thought.

Dinoflagellates can produce sudden "blooms" when they proliferate in multitudes, causing "red tides" and other visible effects in the seas. The sudden blooms can be toxic to other marine life, and through the food chain to humans. One species causes paralytic Shellfish poisoning in tropical regions, extending as far south as Port Moresby in New Guinea. This species was widespread in the Tertiary and was in Sydney Harbour during the Pleistocene Epoch. There is some anxiety that its range is increasing and it may return to Australian waters. The possibility of this happening is strengthened if the warming of the Greenhouse Effect provides the conditions which it prefers further southwards.

Noctiluca, the organism which produces phosphorescence in the sea, is a Dinoflagellate. It has no chlorophyll, and instead of having a pair of flagellae it has a single tentacle.

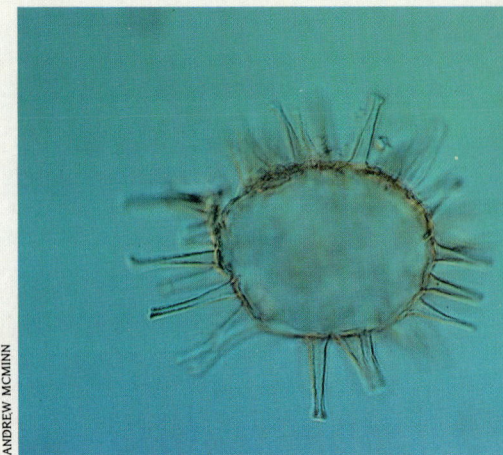

ANDREW MCMINN

Cyst of a Tertiary Dinoflagellate of the same kind as the living one which causes Shellfish paralysis in humans.

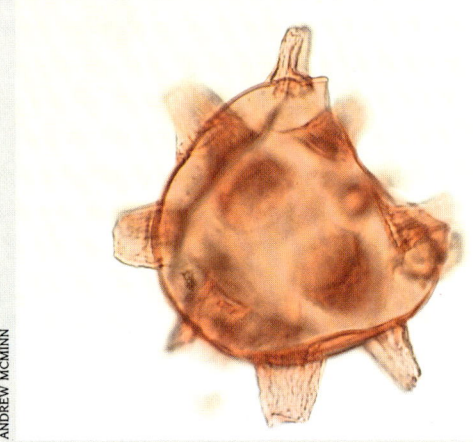

ANDREW MCMINN

A Mid Cretaceous cyst with a plate missing.

ANDREW MCMINN

An Early Cretaceous cyst which has lost a plate.

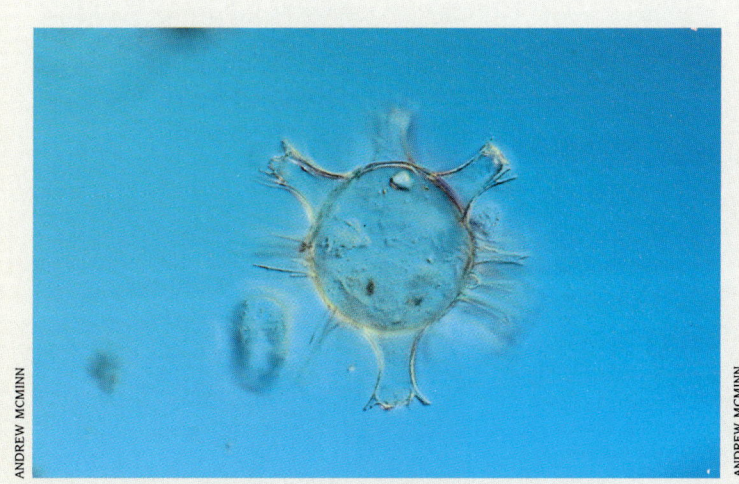

A Miocene cyst.

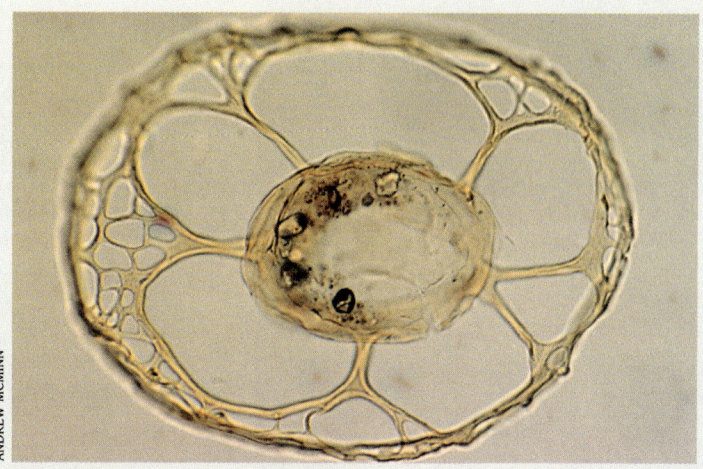

A Mid Cretaceous cyst. The organism has contracted within the cyst wall.

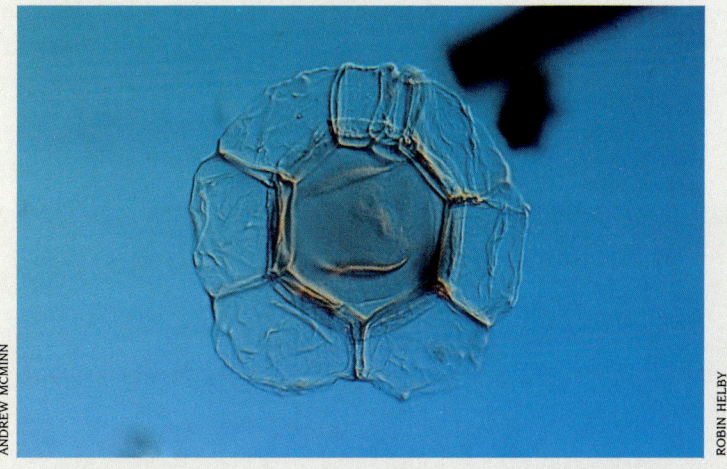

A Miocene cyst.

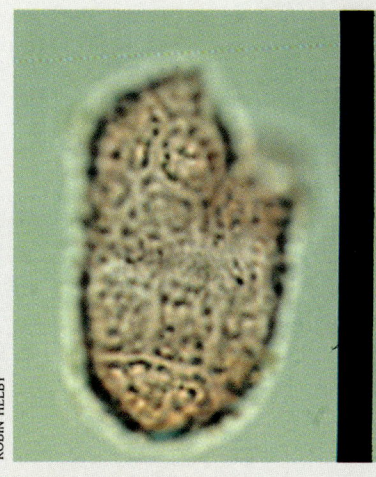

Pyxidiella pandora, a Jurassic Dinoflagellate cyst.

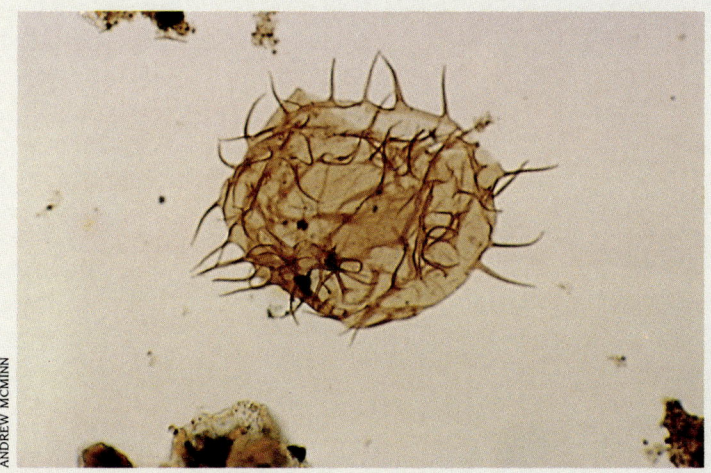

Cyst of a living Dinoflagellate, from Port Stephens, New South Wales.

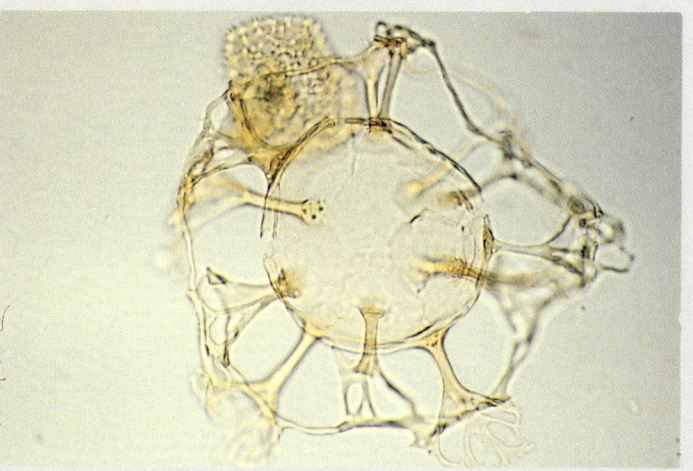

Rigaudella aemula, a Jurassic cyst.

*A JURASSIC BENNETTITALEAN
CYCADOPHYTE FROM LUNE RIVER,
TASMANIA*

*The Bennettitalean ''polycarp'', a many-
seeded fruiting structure characteristic of
this extinct group of Cycadophytes.
(Specimen AMF. Magn.X 1.6)*

OPPOSITE PAGE:
*A median longitudinal section through the
polycarp shows a central receptacle and a
rind of seeds separated by sterile scales.
The white outline of the polycarp is the
result of weathering of the specimen and is
not structural. (Magn.X 4.9)*

RIGHT:
*A thin section of part of the seed layer,
much enlarged. The two cotyledons of the
seeds and the extension of the micropyles
(fertilisation tubes) to the outside of the
''fruit'' can be clearly seen.*

BELOW:
A developing seed, much magnified.

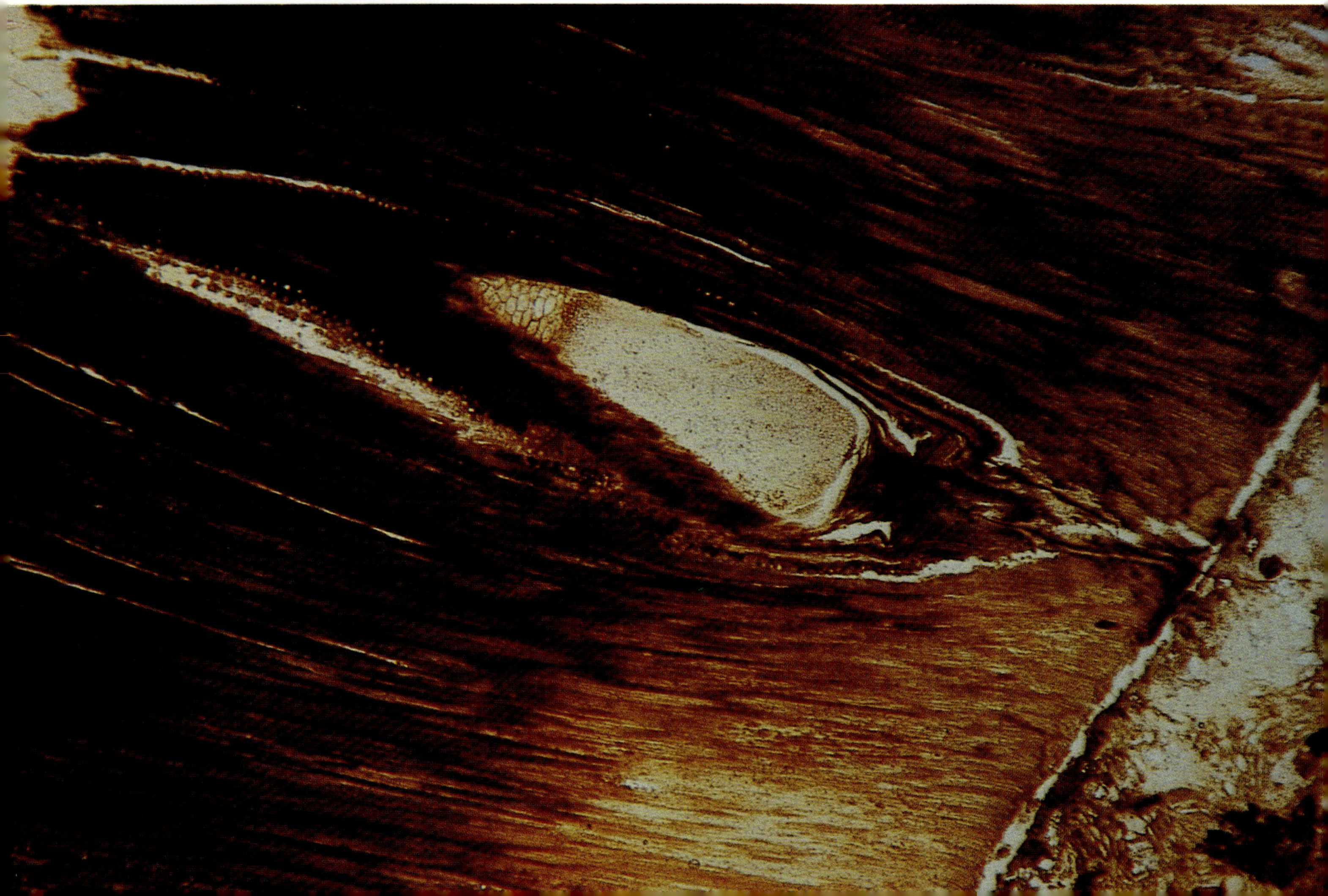

CHAPTER 8

A TIME OF CHANGE

THE CRETACEOUS PERIOD
FROM 144 TO 66.4 MILLION YEARS AGO
DURATION: 76.6 MILLION YEARS

During the Cretaceous Period the split-up of supercontinents gathered pace. The movement of landmasses progressively changed circulation patterns and altered world climates. From this time on, the world was heading towards an ice age. Australia remained attached to Antarctica and in high latitudes throughout the Period, while Ancestral New Zealand remained attached to Marie Byrd Land until the Late Cretaceous.

Major flooding occurred worldwide in the Early Cretaceous. In Australia there was a vast inland sea. Ancient-style vegetation dominated by Gymnosperms thrived under the warm to hot and wet conditions of most of the Period. Flowering Plants appeared in the Late Cretaceous and started to spread and to invade the old-style communities, in preparation for their rise to dominance in the Tertiary.

Dinosaurs, Pterosaurs and marine Reptiles were abundant for most of the Cretaceous, but declined rapidly towards the end of the period and largely became extinct. Australia's oldest Mammal fossil, an opalised jaw of a Monotreme some 110 million years old, indicates for Australia the start of the rise of the Mammals.

Dinoflagellates and Foraminifera were abundant in plankton and there was a rich marine Invertebrate fauna.

The end of the Period saw many extinctions. Along with Dinosaurs and marine Reptiles among the Vertebrates, Ammonites and many other Invertebrates disappeared. Modern-aspect faunas were effectively established after this extinction event.

A Cretaceous Ammonite with a partially uncoiled shell. Most Ammonites have closely coiled shells, and an atypical example of the group, like this one, is known as a heteromorph. It was formerly thought that aberrant forms were the result of "racial senescence" and that they appeared just prior to the extinction of a group. This view is no longer generally held, as heteromorphic shells appeared at different times and there is no direct relation to the time of extinction. (Specimen MQU. Magn.X 3.0)

An Ammonite from the Cretaceous of Queensland. The complex sutures can be seen on part of the shell. Ammonites range from the Devonian to the Cretaceous, and are abundant and useful in determining zones in Mesozoic faunas. (Specimen U.NSW. Magn.X 0.9)

Early Cretaceous Bivalve Molluscs, Pseudavicula anomala from Maranoa, Queensland. (Specimen MMF. Magn.X 1.3)

The Cretaceous Period is aptly called a "time of change". Dramatic events on the world scene — with rifting and drifting, mountain-building and flooding, and altered climatic patterns — meant this change was on a vast scale. The Period encompassed the transition from ancient-style vegetation towards flora dominated by Flowering Plants and from the reign of the Reptiles to the start of the rise to dominance of the Mammals.

The plate tectonic movements which were starting to break the supercontinents asunder resulted in a global sea level rise in the Early Cretaceous. Very large areas of land surface in all parts of the world were flooded. The sea level reached its maximum height about 110 million years ago and then fell rapidly. By the Late Cretaceous most of the land was above the sea again. A new round of mountain-building associated with plate movements established the Rockies, the Andes and the Alps.

At the start of the Cretaceous there was a sudden cooling episode. How severe it was and how long it lasted are unknown. However, throughout the rest of the Period temperatures rose steadily and for the most part the Cretaceous was a time of great warmth with mean annual temperatures between 10° and 15°C warmer than they are today and the temperature gradient from the Equator to the Poles about half of what it is now. Tropical and subtropical conditions extended far further north and south than at present, possibly up to the 70° parallels at times, and there was a warm-temperate climate at the Poles. Abyssal water is estimated to have been at 15° (it is 2°C today), ocean circulation was sluggish and in consequence vertical zonation was almost absent.

Widespread evaporite deposits are found in a band extending to about 45° on either side of the Equator, indicating that there were times of aridity.

The Late Cretaceous was the warmest time interval in the Phanerozoic Eon. However, a sudden cooling episode at the end of the Period heralded the onset of fluctuating climatic patterns leading up to the most recent ice age.

During the time when the sea level was rising, there was fluctuation in the amount of land which was flooded locally in areas where tectonic movements were raising and lowering blocks of crust. Thus, even before the epicontinental seas began to retreat, new local environments were waiting to be colonised by plants. They offered new opportunities for plants capable of adapting to the salt marshes, swamps, dunes, seabed sands and the other disturbed areas which appeared following the changes in sea level. In the major global regression that was to follow, it was the opportunity offered on a large and general scale by the emerging new environments which was to act as a spur to a burst of evolution. The experimental Flowering Plants which had been adapting to the changing conditions were ready to diversify and take over.

The timing of the very important evolutionary advance in the Plant Kingdom, the attaining of Angiospermy, was crucial in the subsequent development of modern-style world vegetation. (Angiospermy involves: the enclosing of the seed or seeds in a vessel; double fertilisation resulting in the supply of extra food, or endosperm, for the developing embryo in the ovule; and the evolution of flowers.) The origin of ancestral Flowering Plants before the fragmentation of the supercontinents had been accomplished enabled the ancestors of today's main groups to spread to all lands. The evidence for this early stage in evolution of Flowering Plants comes from the microscopic fossil record, the Pollen Record. The study of pollen, Palynology, is the tool mainly used to understand the history of the Monocotyledons and Dicotyledons, the two main sections of the Angiosperms.

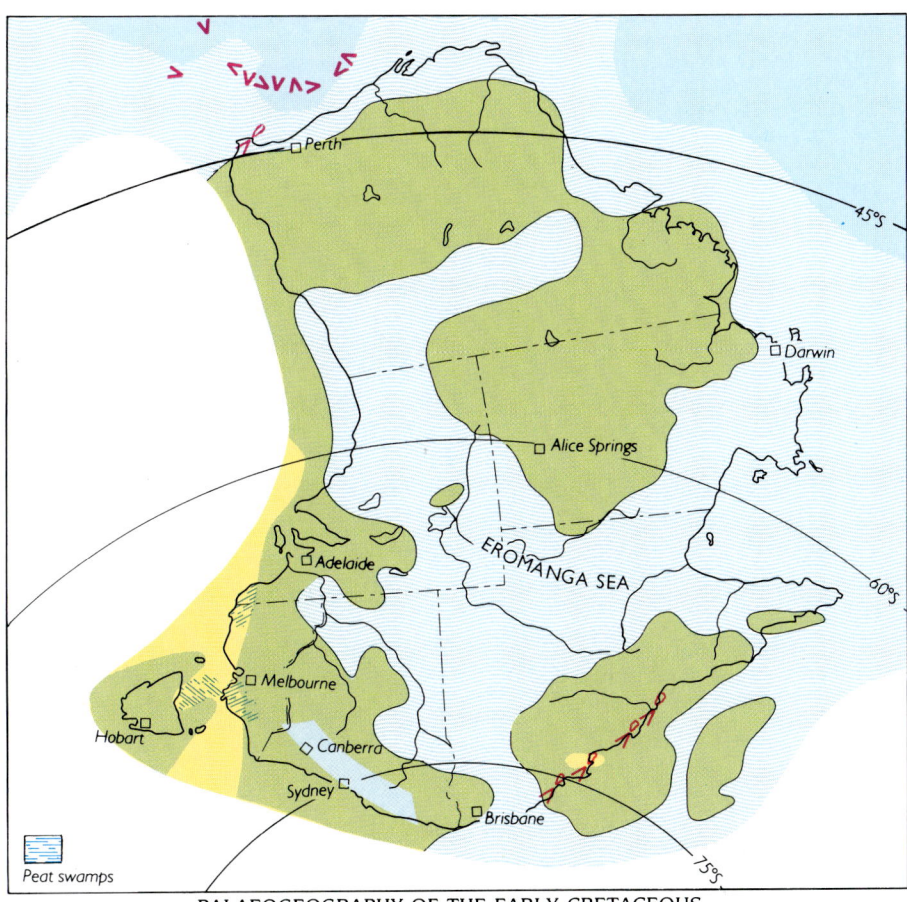

PALAEOGEOGRAPHY OF THE EARLY CRETACEOUS
120 million years ago

Peat swamps

EROMANGA SEA

PALAEOGEOGRAPHY OF THE LATE CRETACEOUS
80 million years ago

GREAT DIVIDE

Gantheaume Point, an important fossil locality for Dinosaur footprints.

Dinosaur footprints in the Broome Sandstone on Gantheaume Point, Western Australia.

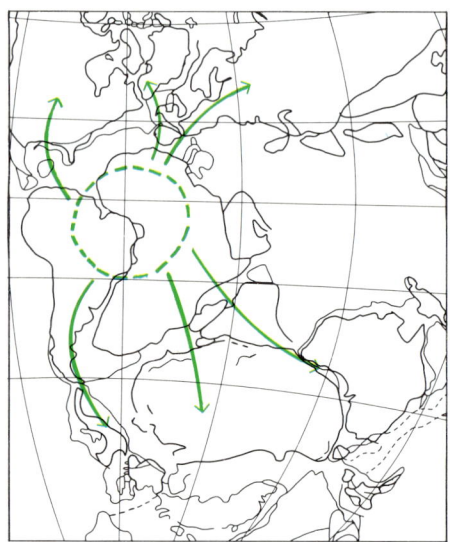

RADIATION OF EARLY ANGIOSPERMS
Early Cretaceous Reassembly, showing West Gondwana as a centre from which the first Flowering Plants could spread North and South.

A magnified view of sutures in the Late Cretaceous Ammonite, Pachydiscus, from the Miria Formation of Exmouth Gulf, Western Australia. (Specimen WAM.)

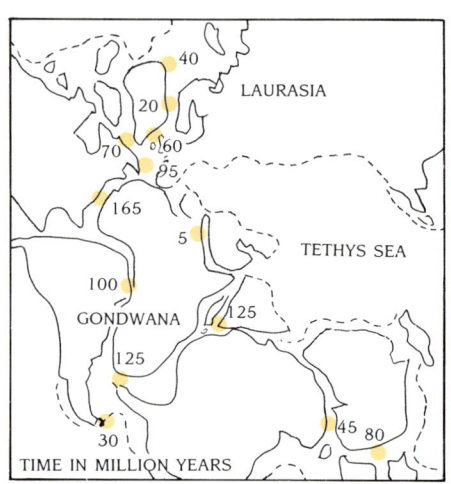

TIMES OF SPLIT-UP OF PANGAEA

The oldest Angiosperm pollen that has been recorded is about 124 million years old (from the Barremian division of the Cretaceous) and comes from localities in Laurasia and in western Gondwana. Following on the appearance of this first "generalist" pollen, primitive pollen of Magnolioid, Lauralian and Lilioid types is found in rocks which were formed during the next few million years, showing that an evolutionary radiation was under way. A concentration of early pollen types in rocks of about 120 million years old in western Gondwana suggests that this region was a major centre for Angiosperm evolution. Rapid diversification and the establishment of many of the families of Flowering Plants occurred within a relatively short time. From this locality they could spread widely into both Hemispheres.

The western Gondwanan distribution centre was situated in an area where a vast rift valley was developing between northern South America and North Africa.

Many of the early Flowering Plants may have been well suited to the disturbed conditions in rift valleys, where swamps, sea encroachments and altered soil conditions formed part of the processes of sedimentation and gradual drowning. Recent analysis of the occurrences of earliest Angiosperms seems to indicate that the places in which they lived were seasonally arid. They probably initially occupied riverine and fringe environments and progressed to be "weed" shrubs and spindly trees infiltrating the established vegetation.

Big rift valleys have a microclimate insulating them from the climatic belt in which they lie. The modern rift valley in Africa, with its marvellous prehistoric-looking flora of giant groundsels and *Lobelia* in the Mountains Of The Moon, gives us an idea of what the ancient rift valleys that were forming on lines of separation between continents were like. The network of rift valleys on the supercontinents at the time when the Flowering Plants were evolving not only explains their origin (because of the new environments they provided early in the rifting), but it also explains their diversity since a variety of habitats is always enclosed in a well-evolved rift valley. Early dispersion within the microclimate of such a rift system would also have been possible.

There are diverse reasons for the rise to dominance of the Flowering Plants and the relegating of the Conifers, Cycads, Ferns and others to less significant status in most ecosystems. The ancient-style vegetation of the Late Jurassic had shown some signs of loss of diversity. The more efficient Flowering Plants had the required genetic potential to change to suit all the new niches and to adapt to all the changing circumstances. They had the energy of youth in comparison to the mainly slow growing, slow maturing, individually long-lived plants which they were preparing to replace. They had better seeds which gave their offspring a better chance of surviving, and they had better distribution mechanisms. Rapidly evolving interrelationships between Flowering Plants and Insects, and also between Flowering Plants and Mammals and Birds, led to the development of sophisticated pollination mechanisms which gave the Angiosperms a competitive edge. With these better pollination mechanisms and more efficient agents, there was more rapid evolution. New varieties were created and were available to be selected when changed conditions favoured a new set of adaptations.

The timing of the break-up of the supercontinents during the period when the Flowering Plants were evolving determined the spread of the major plant groups and families to the parts of the world where they are found today. The distribution of Mammals was also determined by the sequence of events in continental dispersion. Placental Mammals which would soon assume dominance in most parts of the world did not reach Australia or New Zealand before they separated from other lands, leaving the way clear

EARLIEST POLLEN OF FLOWERING PLANTS FOUND IN AUSTRALIA

Sites in the Great Artesian Basin have yielded pollen samples of Albian age, about 110 million years old. The simple pollen grains cannot be assigned to any living Angiosperm family. They are classified according to the nature and arrangement of their pores (colpi) or grooves (sulci).

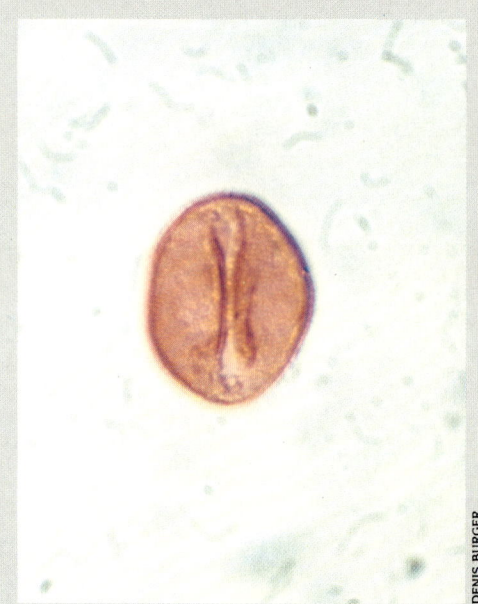

Clavatipollenites, a monosulcate grain.

Rousea, a tricolpate grain.

Phimopollenites, a tricolpate grain.

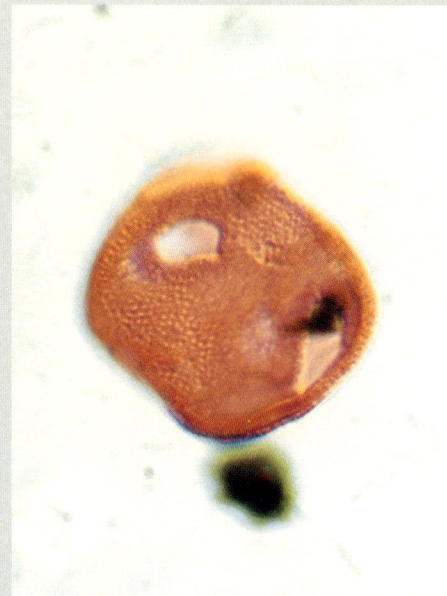

Hammenia, a stephanocolpate grain.

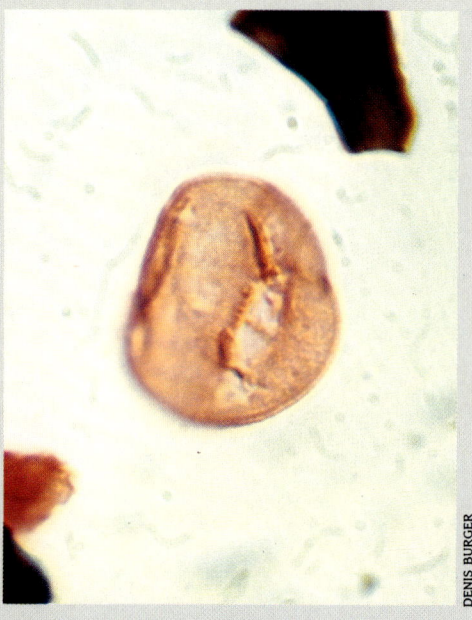

An unidentified tricolpate grain.

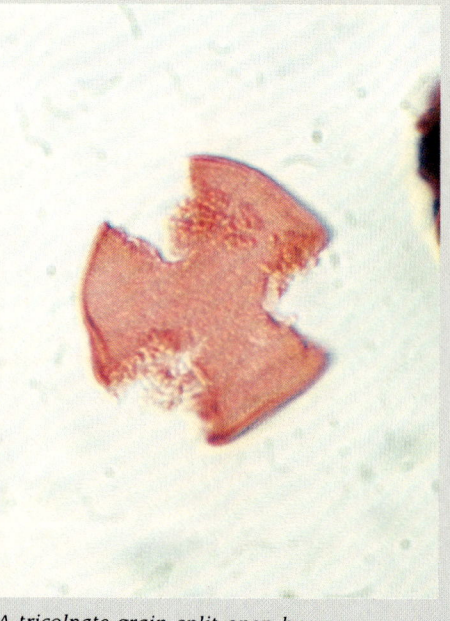

A tricolpate grain split open by compression.

for the unique rise of the Marsupials in Australia (and of large Running Birds in New Zealand in due course). During the Cretaceous the Marsupials were an almost invisible component of the fauna, for this was still a time when the Reptiles reigned.

Two controversial issues relating to Dinosaurs have been the subject of debate: firstly, Dinosaurs have been held responsible for the rise of the Flowering Plants, and secondly, the Flowering Plants have been blamed for the demise of the Dinosaurs.

It is impossible to assess whether or not the decline of the old-style vegetation was contributed to by the relentless grazing of herbivorous Dinosaurs. There were Dinosaurs of all kinds, from monstrous Sauropods to small Bird-like ones, and all sizes between. The huge kind with strong back legs, and reduced front ones, were "tripedal". They sat up on their haunches to graze and were able to eat the delicate growing-tips of large

Nothofagus woodland with emergent Araucarian Conifers in South America.

159

A skull of *Leaellynasaura amicagraphica, a small Dinosaur with large eye sockets and an enlarged optic section of the brain — adaptations for night vision under the conditions which prevail in polar regions where winter brings several months of darkness.*

GREAT ARTESIAN BASIN
Extent of Cretaceous Sediments

trees some 15 metres above the ground. They needed massive amounts of fodder and had gizzards full of well-worn stones to grind the indigestible food. They probably had several large stomachs acting as digesting vats to break down the lignin and cellulose.

In high latitudes the plants were dormant during the winter months because of polar darkness. However, they remained a continuing food supply throughout the dark times, and consequently may have suffered some overgrazing stress. Reduction of plant cover or damage to vegetation by local overgrazing would have offered opportunities for "weeds" to invade established vegetation.

Whether or not the vegetation change to a modern-style Broad-leaf type was a factor which contributed to the decline and demise of the Dinosaurs is also a question which cannot be answered. It may be that some Dinosaurs were so highly adapted to the ancient plant diet that their guts and digestive processes could not cope with the bloat-producing Broad-leaf fodder containing new alkaloids and compounds.

AUSTRALIA IN THE CRETACEOUS PERIOD

During the Cretaceous Period Australia remained attached to Antarctica, though a complex system of rift valleys was developing between the two landmasses in preparation for separation. Latest evidence suggests that Australia's high latitude situation may have been even more extreme than had been previously estimated. Instead of lying between 40° and 75° South, it may have been between 50° and 85° South.

The rising sea levels of the first 50 million years of the Cretaceous resulted in the flooding of the Tasman Depression (as it continued to founder) and the adjacent low-lying central and southern basins, as well as the basins in Western Australia where the Amadeus Transverse Zone used to be. The vast inland waters of the Eromanga Sea separated the landmass into four blocks.

Sedimentary deposits of Cretaceous age are present in outcrop or subsurface over approximately a third of the present Australian continent, due principally to widespread marine deposition in the epicontinental seas. Terrestrial deposits which formed at the start of the Period before flooding was major, and towards the end of the Period when seas had retreated, are less well represented.

As well as the generally gradual and small-scale tectonic unrest caused by events associated with continental break-up, there was a major mountain-building episode in the east coastal region of Queensland. This Maryburian Orogeny deformed the area and was the final act within the Tasman Orogen.

On Australia's southern margin, thick sediments continued to accumulate throughout the Cretaceous in basins which were formed by the rifting that preceded separation from Antarctica. In Bass Strait these sediments contained coal. The rifts of the Gippsland, Otway and Bass basins evolved slowly, receiving continental sediments and volcanic detritus. By late in the Period, sea invaded the rifting zone progressively from the west until Australia was connected to Antarctica only by a trailing edge through Tasmania.

The split-up of the Gondwanan supercontinent gathered pace throughout the Cretaceous. Australia's western margin was progressively delineated. Seafloor spreading first occurred near Carnarvon and Perth along the edge of the Westralian Depression, and major sequences of source and reservoir rocks were laid down in the Barrow Island oilfield in the Carnarvon Basin. At the same time India was freeing itself from the western arm of the Depression. India's movement northwards started at about 125 million years ago. Simultaneously Africa rifted from Antarctica and started to rotate away from India.

Australia's south-eastern margin was established by the start of the opening of the Tasman Sea some 80 million years ago. The Lord Howe Rise and the Ancestral New Zealand landmass — extending from the Campbell Plateau to New Caledonia and comprising the whole outer edge of Gondwana — started moving away from what is now Australia's east coast. About 60 million years ago the movement was complete and the distance between Australia and New Zealand has been constant ever since. The tilting of the eastern margin of Australia and the formation of the Great Divide resulted from the tectonic events involved in the opening of the Tasman Sea.

Antarctica had come to straddle the South Pole by the end of the Cretaceous (and has remained there) while the lands which were attached to it moved steadily northwards — or, in the case of Australia, prepared to do so.

The climate in Australia during Cretaceous times was very warm and humid, apart from a cold spell at the beginning of the Period and another at the end. The cooling episode in the Early Cretaceous is evidenced by the "winter kill" of Fish in the Koonwarra Fish Beds in Victoria and by ice-rafted boulders among the sediments in the Eromanga Sea. Winters must have been cold during this time and there may have been glaciers on high ground. Marine plankton studies also indicate that seas were cool to cold off south-eastern Australia at the start of the Period. This cooling episode would have resulted in environmental stress on the vegetation, which had evolved and thrived under the uniformly warm and humid

The larva of a Mayfly from the Koonwarra Fish Beds in Victoria. (Specimen AMF. Magn.X 1.6)

DENSEY CLYNE

A living Mayfly.

OVERLEAF:

AN EARLY CRETACEOUS LANDSCAPE — DINOSAURS AT THE SOUTH POLE

120 million years ago

The uniquely Australian *Muttaburrasaurus* and various small bipedal Dinosaurs were part of the fauna in high latitudes during these warm and humid times. Pterosaurs (flying Dinosaurs) were the notable flying creatures of the times, though the first Birds had appeared (as evidenced by feathers) and would shortly dominate in the skies. The vegetation of Conifers, Cycads and Ferns was luxuriant. Australia was a "land of the midnight Sun", where in mid-summer there was no night. It was also a land of long dark winters when there was no daylight, although the temperatures remained warm. *Pentoxylon* Cycads with their red fruits are seen in the right foreground and Araucarian (Southern) Conifers at top right. Cycads, Tree-ferns and Ferns were abundant. The picture captures the golden light of a long summer "night" when the low-angle of the Sun imparted a special richness to colours.

AN EARLY CRETACEOUS LANDSCAPE —
DINOSAURS AT THE SOUTH POLE

A landscape in the Mitchell River region of Queensland in which horizontal rock strata of Cretaceous age cap Permian volcanic rocks. The plain comprises folded Silurian and Devonian sediments.

conditions of the Jurassic. Further stress was to be added by the gradual drowning of the land as the sea level rose and the epicontinental sea expanded.

The Fossil Pollen Record shows that early "generalist" Angiosperm pollen appears in Australia in Albian sediments from 110 million years ago, which is about 14 million years later than it appeared in West Gondwana and Laurasia. Near the end of the Late Cretaceous, pollen which can be identified as *Ilex* (Holly), Proteaceous-type and *Nothofagus brassii* (a southern Beech) make their appearance in the Record. Vegetation was still dominated by Conifers (at least in the few areas from which there are pollen samples). Several modern Conifer types were already abundant. Among the Podocarp pollen, Huon Pine, Rimu (the common New Zealand species which is also found in Tasmania) and *Phyllocladus* (the Celery Top Pine) are present. Araucarians with pollen like that produced by *Agathis* (Kauri), Hoop, Bunya and Norfolk Pines were abundant. Pollen of Cypress Pine like modern *Callitris* has also been identified from Cretaceous times.

The Fossil Pollen Record and the modern distribution pattern of *Nothofagus* shows that this genus of southern Beech originated after the Africa–India block of Gondwana was effectively isolated, as it is absent from those lands. It spread to Australia and to the New Zealand to New Caledonia outer edge of Gondwana at a time when they were still joined to Antarctica and South America. *Ilex* is an old genus with cosmopolitan distribution, having been established before fragmentation of the supercontinents could interfere with its dispersion into Laurasia and Gondwana. Now it has all but disappeared from Australia, leaving one species in Arnhem Land in the Northern Territory. There are several species

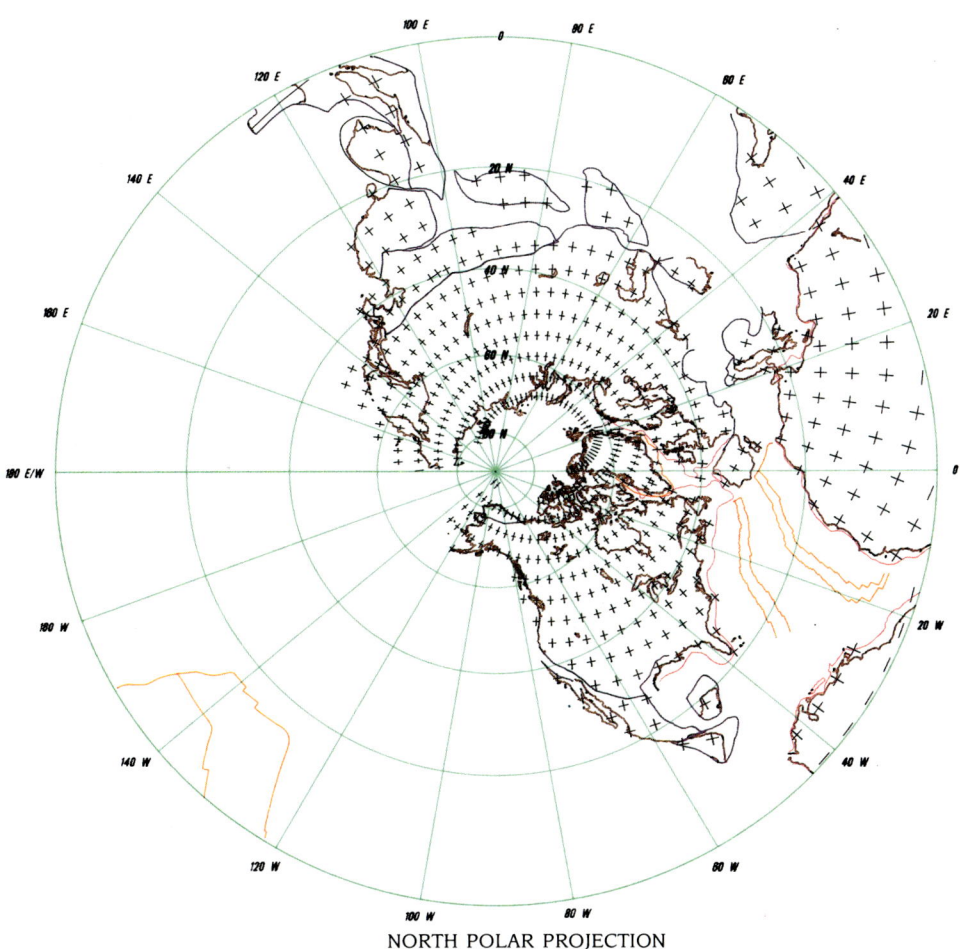

NORTH POLAR PROJECTION

Odontochitina porifera.

ROBIN HELBY

Muderongia australis.

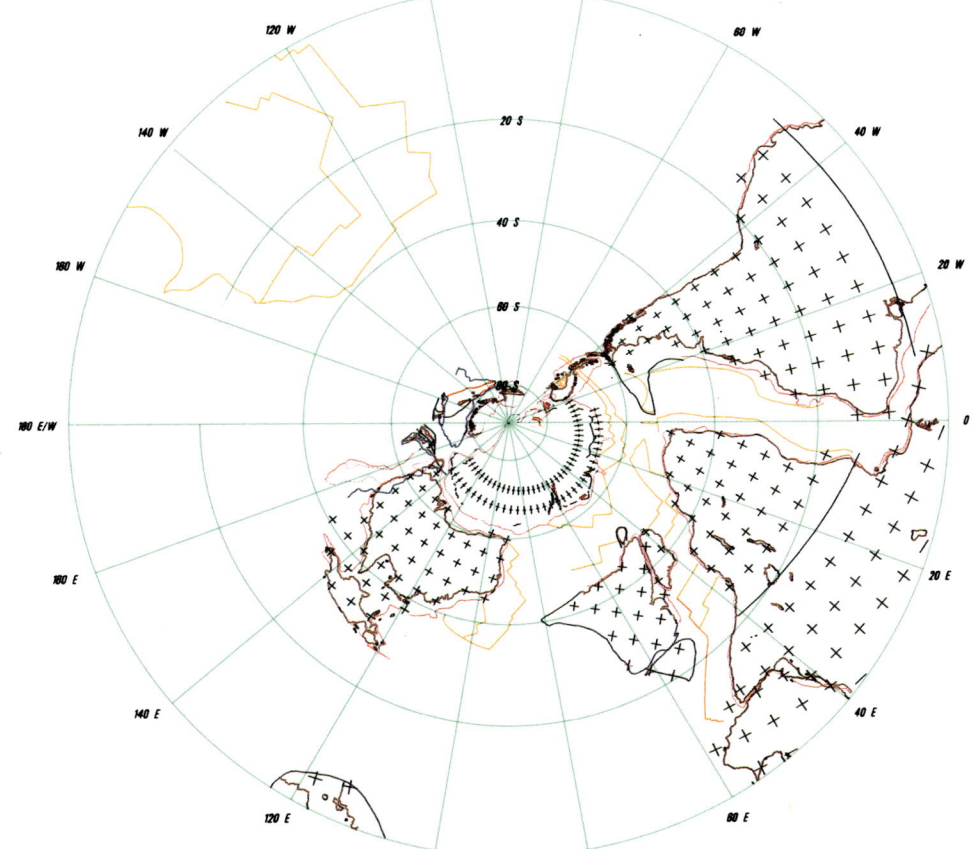

SOUTH POLAR PROJECTION

*POSITION OF LANDMASSES
IN RELATION TO THE
NORTH AND SOUTH POLES
IN THE CRETACEOUS PERIOD,
100 million years ago*

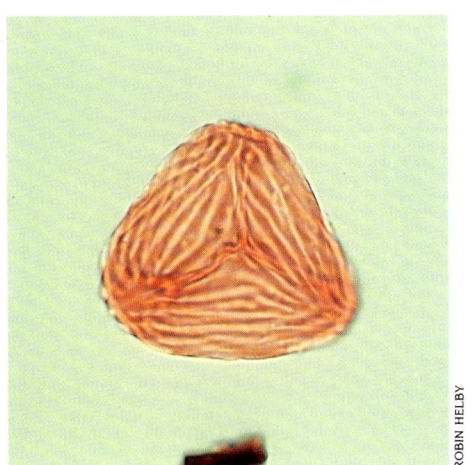

Cicatricosisporites australiensis, a microspore.

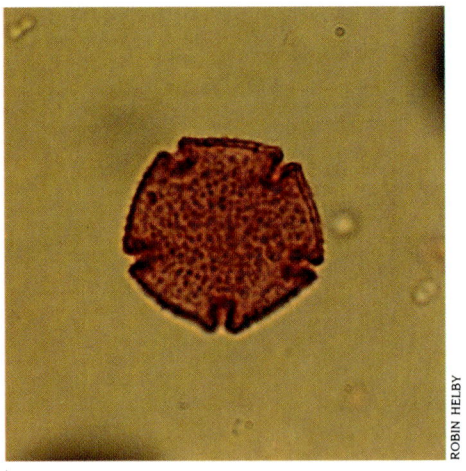

Nothofagidites senectus, pollen of southern Beech.

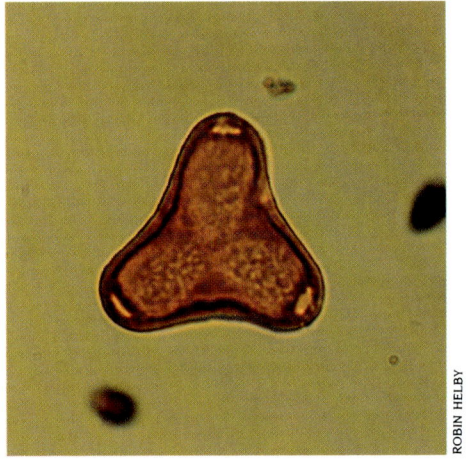

Triporopollenites sectililis, a pollen grain.

today in New Guinea. *Ilex* is an example of a genus of plants which migrated into suitable refuges in the Cainzoic Era when aridification of Australia caused many Rainforest plants to contract their ranges. The PROTEACEAE is an ancient Gondwanan family which spread into all southern lands while they were still connected.

Early Cretaceous macrofossil floras of Victoria are rich and varied assemblages comprising Conifers, Ginkgophytes, Pentoxylean and Bennettitalean Cycadophytes, Tree-ferns, Ferns and Seed-ferns. These plants are of the same groups which characterised the Jurassic. The abundance of Filmy Ferns and Mosses indicates cooler, moist conditions. The similarity between the Early Cretaceous Victorian flora and that of Alexander Island in west Antarctica is remarkable. A Northern Territory flora, also of the same age, has a great preponderance of Cycadophytes which seems to indicate that the climate there was hotter and drier.

South America has significant Cretaceous macrofloras comprising "subtropical" Broad-leaf Angiosperms and "cool-temperate" Podocarps and *Nothofagus*. Such floras are described as "mixed floras" because of the modern preferences of their living relatives. No macroflora of this sort has yet been discovered in Cretaceous rocks of Australia. New Zealand, however, has an incredibly rich Early Cretaceous flora which is currently being studied at Christchurch University. The presence of this New Zealand flora implies that there must be a gap in the Fossil Record in Australia, which may in time be filled.

The use of the term "mixed flora" is very misleading. The conditions under which the Antarctic or "cool-temperate" element and the "subtropical" element had evolved throughout the Mesozoic were warm-temperate. It is important to remember that it was warm and humid at the Pole when this vegetation type was widespread.

The Fossil Record of animal life in the Cretaceous in Australia is rich and diverse. It includes Dinosaurs of many sorts and sizes, Pterosaurs (flying Reptiles) and an assortment of Dinosaurs track. In the sea were other large Reptiles — Turtles, Ichthyosaurs and Plesiosaurs.

The Winton Formation in Queensland contains a famous rock platform where footprints of Dinosaurs tell the story of a stampede of small Dinosaurs fleeing from a large Theropod Dinosaur which probably disturbed them at a waterhole. In the Otways in Victoria medium and small sized Dinosaurs are being found in increasing numbers at Dinosaur Cove. Some have enlarged eye sockets as well as enlargements in that part of the brain concerned with vision — adaptations for seeing better in the dark in the high latitudes in which they lived.

Also in Victoria, the Koonwarra Fish Beds have yielded four feathers, which are the first evidence of Birds in Australia. Birds evolved from Dinosaur stock. Fleas have also been found associated with the plants and Fish in this formation. Fleas have been recorded with Pterosaur fossils in Russia, so ancient Cretaceous animals were already plagued by parasites of a modern kind.

A uniquely Australian Dinosaur, the *Muttaburrasaurus*, is known from an almost complete skeleton found in the Early Cretaceous Mackunda Formation near the small town of Muttaburra in Queensland. It was a large Bird-footed and Bird-hipped creature with an enlarged bony chamber on its nose, and its front teeth were fused into a beak which it used for tearing off vegetation as it grazed. Remains of *Minmi*, an armoured Dinosaur, and of a huge Sauropod and a big predatory Theropod similar to *Allosaurus* have also been found in Queensland.

At Lightning Ridge in western New South Wales, Lungfish like those alive in Queensland today occur as fossils. During the Early Cretaceous the

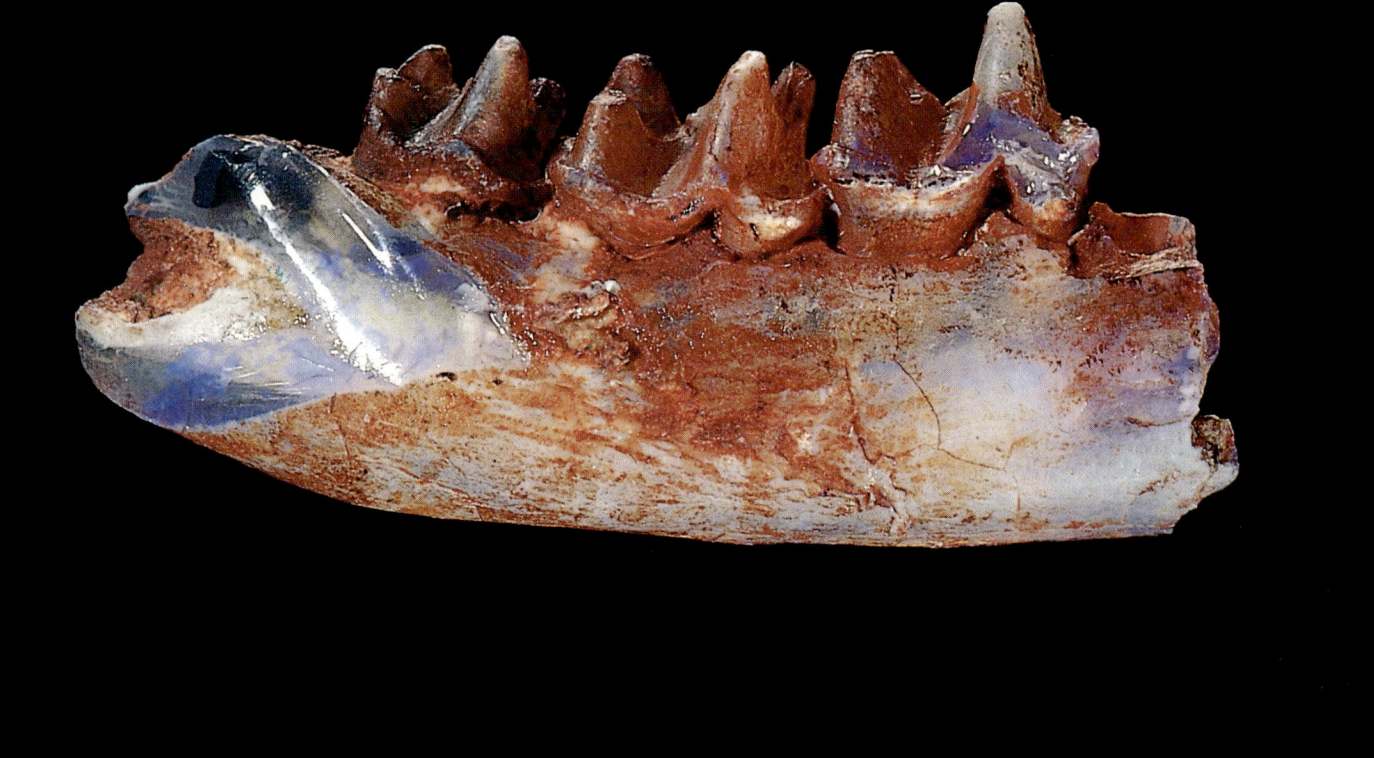

region where Lightning Ridge now stands was an estuarine area where a major river flowed into the inland sea — a landscape very different from the waterless moon-landscape place it is today. Opalised Plesiosaur fossils have been found there: some are almost complete skeletons.

The most significant Vertebrate fossil find in Australia, and arguably one of the most important in the world, was made at Lightning Ridge. It is a small opalised jaw, 110 million years old, which was bought in 1985 by the Australian Museum in Sydney as part of the Galman Collection of opalised fossils. Called *Steropodon galmani*, it is the jaw of an early Platypus. It is the oldest Mammal fossil found in Australia to date. (Although Mammals are known to have been around in other parts of the world since Triassic times about 200 million years ago, which is 90 million years before *Steropodon*, they were inconspicuous members of Dinosaur-dominated faunas.)

The rich marine Invertebrate faunas of the Cretaceous Period contain Ammonites, Belemnites, Sea-stars, Crinoids, Brittle Stars, Bivalves, Gastropods, Brachiopods and Sponges. Microplankton were abundant in the shallow seas, and the Dinoflagellates and Foraminifera are useful in determining zones and ages in the Cretaceous marine sediments.

While the Cretaceous Period was a "time of change" which brought the Mesozoic Era to a close, it is also separated from the Cainozoic Era by a very clear dividing line often referred to as the "Terminal Cretaceous Event". On one side of this transition at 65 million years are "prehistoric" landscapes, climates, vegetation and animals. On the other side are easy-to-relate-to modern-style vegetation, familiar-looking animals not so far removed from those of our times, and conditions which become progressively more like the present. A number of phenomena mark the transition, inciting much speculation and theorising about their significance and about the processes involved in the changes.

The opalised jaw of Steropodon galmani, an Early Cretaceous Mammal (a Monotreme). This Platypus jaw, dated from 110 million years ago, is Australia's earliest Mammal fossil. It was found by opal miners at Lightning Ridge and sold to the Australian Museum as part of the Galman Collection. (Specimen AMF. Magn.X 6.1)

Fumarole activity in a volcanic landscape.

A volcano erupting on Iceland.

THE TERMINAL CRETACEOUS EVENT

The boundary between the Mesozoic and Cainozoic Eras in the Geological Time Column is well defined. A number of factors contribute to what has become known as the "Terminal Cretaceous Event" which forms the boundary and separates the prehistoric ages from those of more modern aspect.

The most obvious elements of this event are the presence of an iridium layer in sedimentary rock sequences, the massive extinction of marine microplankton, extinctions of some marine Invertebrates including the total disappearance of the Ammonites, and the apparently sudden extinction of Dinosaurs after their long and successful reign in the Mesozoic Era.

At first glance the extinctions and the deposition of the iridium layer seemed to coincide. It is easy to forget that intervals as short as a few million years do not register in the Fossil Record and cannot be readily demonstrated. Consequently, making an exact correlation of events is a major problem. In terms of the lives of animals, a million years or even 100,000 years is a very long time.

Intensive investigation of the crucial interval of geological time at about 65 million years ago by many experts in different disciplines is throwing light on some aspects of the Terminal Cretaceous Event and dispelling some of the myths. From all the evidence gathered and analysed it emerges that there was no single "event", but rather a period of some duration during which a number of events occurred and certain processes which had been in train for some time were brought to their conclusion.

At about the Cretaceous–Tertiary time boundary there was a sudden sharp fall in sea level. (The sea level worldwide had been high in the Early Cretaceous, causing extensive flooding. It had subsequently fallen, the epicontinental seas had retreated, and the sea level had then stabilised.) Although there was a global cooling episode at this point, there is no direct evidence of ice cap formation. The sea level drop is attributed by some authorities to the subsiding of vast submarine mountain systems in the Pacific Basin as a result of plate tectonic movements. This drop in sea level, resulting in the loss of shallow-water environments, would have affected many marine Invertebrates. The Permian Marine Collapse at the Palaeozoic to Mesozoic transition was similarly contributed to by the disappearance of shallow-water environments.

The iridium layer is a thin clay band, very rich in the rare earth iridium, which appears in geological sequences around the 65-million-year time slot. The layer was first described from Italy and Denmark, and it was argued that such a concentration of iridium could only be explained if an extra-terrestrial source was invoked. The idea was that an asteroid of great size had collided with the Earth, throwing up a vast cloud of dust and particles into the stratosphere. When the dust cloud settled, it formed the iridium-rich clay band. It was further argued that the dust cloud had shut out the sunlight and the vegetation had died, causing the Dinosaurs to starve to death suddenly at this time. This idea caught the imagination of scientists everywhere and in their uncritical enthusiasm iridium layers were detected all over the place and assumed to be from the exact time interval of the Terminal Cretaceous Event.

Since the first enthusiastic acceptance of the idea as a possible explanation for what appeared to be a single event requiring a dramatic explanation, sanity has returned to the debate. It now seems certain that there is no ubiquitous iridium layer, and neither are the known occurrences necessarily contemporaneous. Iridium-rich clay bands span 100,000– to

200,000-year sections of sedimentary sequences. Multiple iridium anomalies characterise sections in America, as reported by the Geological Society Of America in 1986. Evaluation of the quantities of iridium and other elements in the clay layer are not inconsistent with an origin in the Earth's mantle and do not require an extra-terrestrial source. So there is one less myth: not only is there no single layer of the same age everywhere, but the composition of the clays is compatible with an earthly source and no asteroid needs to have been involved.

Experts now agree that the selective nature of the extinctions recorded during the Late Cretaceous and the transition to the Tertiary does not support a model which calls upon the blocking out of sunlight for long enough to kill the vegetation and cause global famine. No other comprehensive scenario solely related to impact with an asteroid has been proposed.

An alternative explanation for such high levels of iridiuim in widespread locations is found in an observed increase in the level of volcanic activity. It is believed that a relatively short (in geological terms) period of intense volcanism, on a global scale, led to volatile emissions, acid rain, reduction in alkalinity of surface ocean waters, global cooling, and ozone depletion. Such a pulse of intensified vulcanism would have created the iridium-rich clays over the period. It would also have caused the termination of some of the processes of decline which were already under way in response to stresses imposed by deteriorating temperature regimes, reduced habitat and other environmental changes. The vulcanism presumably represented a critical stage in plate tectonic events as the continents moved and the seafloors spread.

In the case of the Dinosaurs, the proposition that they became extinct in a dramatic "event" also appears to be a myth. Their disappearance at around 65 million years ago was not so extraordinary when it is seen in context. There had been a series of extinction events in Dinosaur history. A very high turnover rate of species, in the order of a few million years each, is observed throughout the long time when Dinosaurs were conspicuously abundant. New forms arose and replaced each other in a steady process of extinctions and new appearances. The difference at the Cretaceous–Tertiary time boundary was that there was a failure to replace those Dinosaurs which became extinct with new Dinosaurs. In addition, it now appears from recent research on Dinosaur extinction in North America that the slide to extinction took place gradually over about 7 million years preceding the 65-million-year boundary. The process accelerated rapidly in the next 100,000 years, during an interval of apparent competition with immigrant Mammals. The last Dinosaurs in North America are now claimed to be Palaeocene in age and the fossil bones are not derived from the Late Cretaceous strata as previously thought. They appear above the iridium anomaly.

Hence, another myth is being dispersed. There was no sudden and cataclysmic Dinosaur extinction neatly coinciding with the iridium layer, but instead a decline over a period of some millions of years until a compounding of conditions caused their demise. They went out with a whimper, not with a bang.

Globally most groups of Vertebrates survived the Cretaceous–Tertiary transition, including Fish, Amphibians, Turtles, Crocodiles, Lizards and Birds. Among the Mammals, Marsupials were already declining overall and Placentals were increasing in diversity before the Dinosaurs became extinct. The Plant Fossil Record shows no massive extinction, but instead there is a burgeoning of the Flowering Plants.

Plankton extinctions at the Cretaceous–Tertiary boundary may be in part

Tree-ferns (Dicksonia antarctica) and Southern Beech (Nothofagus) in Tasmania. Pollen of Nothofagus was one of the earliest of Flowering Plants to appear in the Pollen Record in Australia.

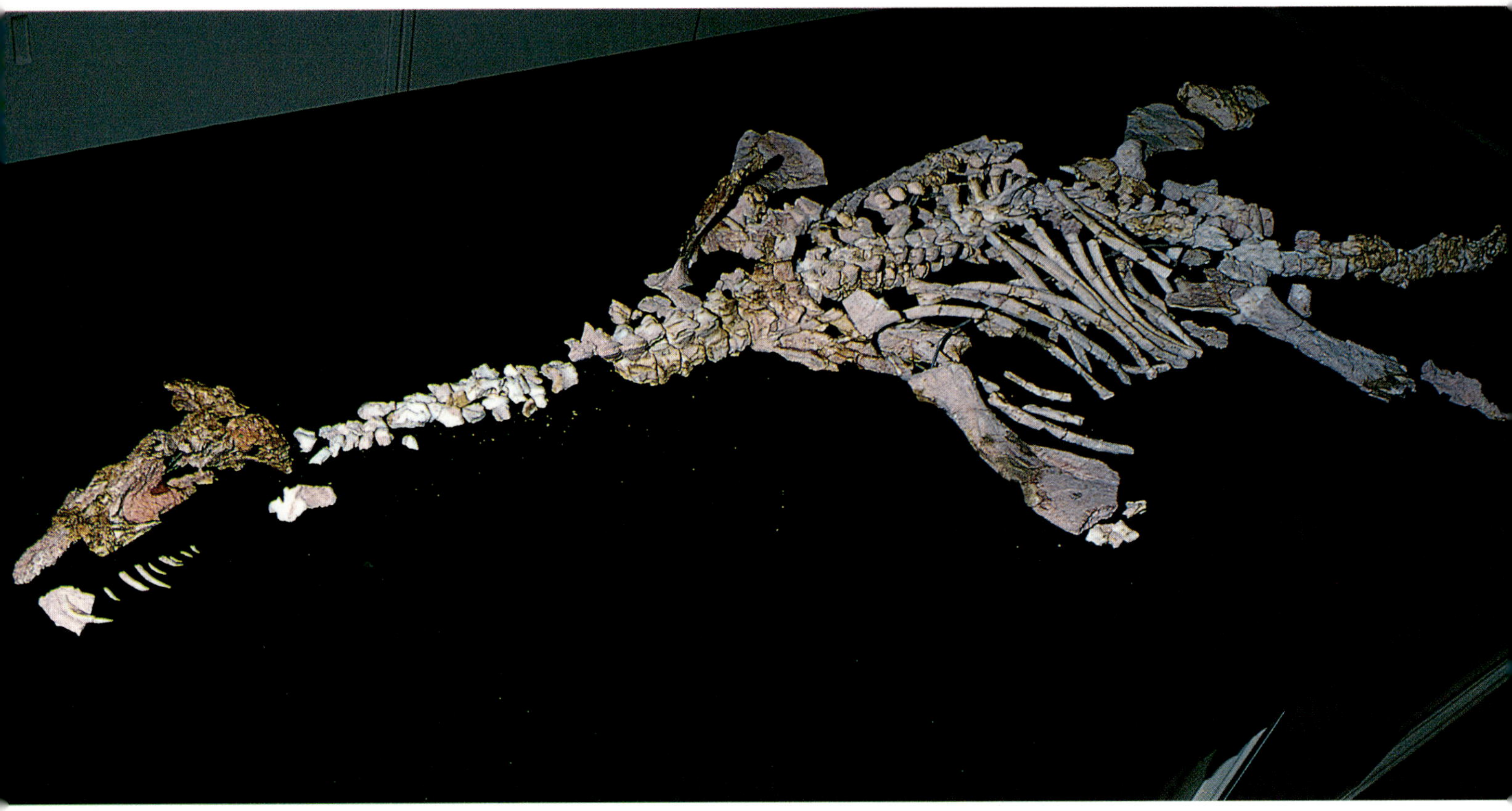

Opalised skeleton of an Early Cretaceous Plesiosaur from Coober Pedy, South Australia. This species is a small Plesiosaur, measuring about two metres from nose to tail. It lived in the island sea which flooded much of Australia at that time. (Specimen Sid Londish.)

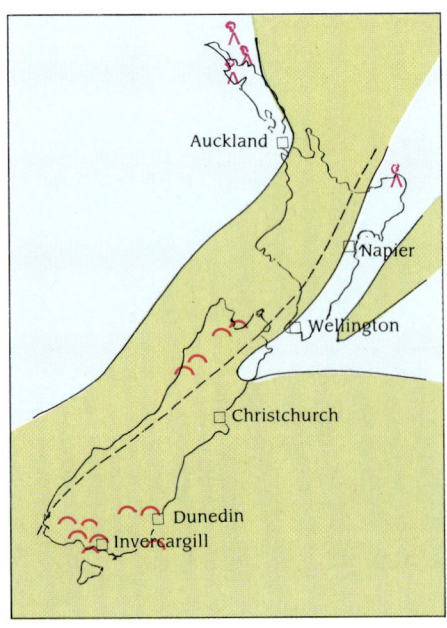

PALAEOGEOGRAPHY OF NEW ZEALAND DURING THE EARLY CRETACEOUS EPOCH

related to the decrease in shallow-water environments. The main cause of their demise is probably explained by the change in acidity of surface ocean waters. Volcanic activity results in emission of sulphur and other elements which create acid rain. The dust and ashes from eruptions, and submarine emissions, have a direct effect on the surface waters of the sea. Planktonic Foraminifera and Nannoplankton (Coccolithophoroids, or Algae with calcareous plates) show different extinction/recovery patterns at the boundary. Both are considerably reduced at the time of the iridium anomalies, but the Nannoplankton continued to experience extinctions well into the Tertiary.

NEW ZEALAND IN THE CRETACEOUS PERIOD

The tectonic movements which had raised the Ancestral New Zealand landmass above the sea in the Mid Jurassic intensified in the Late Jurassic to earliest Cretaceous times, when the Rangitata Orogeny created high ground and the landmass attained its maximum known extent. (There were mountainous regions along the centre of the South Island sector at this time.) The exact size and configuration of the land is uncertain. The shape of the highest part of the landmass in general followed an orocline, a bend in the plate margin over which it lay. (Modern New Zealand still shows the S-shape and straddles the margin between the Australian and the Pacific plates.)

Rapid river erosion reduced the height of the mountains and uplands which had been created in the Rangitata Orogeny, and the extent of the land was further reduced by the progressive sea level rise which affected all parts of the world during the Early Cretaceous. Thus, by about 110 million years ago all of the Northland region was under the sea as were the eastern sides of North Island and of Marlborough in South Island.

Volcanoes were active off the north-eastern and north-western flanks of the land which stretched away northwards to New Caledonia, perhaps more as a chain of islands than as continuous land. (At this time Australia was experiencing its greatest invasion of epicontinental seas, *Steropodon* the early Platypus was living in New South Wales, and the Flowering Plants were evolving and spreading across the still-connected lands. Only the Indian sector of Gondwana had cleared its margins and started to move north on its own.)

During the Late Cretaceous there was rifting between the south-western end of the Ancestral New Zealand landmass and Antarctica, in preparation for the separation of the outer edge of that part of Gondwana. The opening of the Tasman Sea with the generation of new seafloor, between 80 and 60 million years ago, resulted in the Lord Howe Rise and the land to its east moving away progressively from the rest of Gondwana. While the Ancestral New Zealand landmass was moving away, erosion of the land and encroachment of the sea continued relentlessly. By the end of the Cretaceous, at 65 million years ago, the area of the land had been reduced and was only a little larger than present-day New Zealand. The eastern margins of North Island and South Island were flooded and volcanoes were active to the east of the mid region. The land was by now all low-lying and swamps were features of landscapes. Large areas of coal swamp, where plant material produced the carbonaceous matter for lignite and black coal in commercial quantities, occurred in eastern Otago, north-western Nelson and on the west coast (though most of the west coast coals date from later, Eocene, times).

During the time when Ancestral New Zealand remained attached to the outer edge of Gondwana, there was no barrier to migration and spreading of plants and animals. While the land was rising from the sea during the previous Jurassic Period it was being clothed in vegetation, and it had acquired a quota of the populations of the times. It must be assumed that the flora and fauna which inhabited the land at the time of its separation in the Cretaceous were those characteristic of their time, and similar to flora and fauna of other southern lands in high latitudes.

Because the distance which separated Ancestral New Zealand from Antarctica increased progressively from 80 million years ago, the biota which it took with it was a purely Cretaceous assemblage of plants and animals. It was too distant from other lands to receive any younger ones. From this Gondwanan ancestral stock there was to be evolution-in-isolation throughout the Cainozoic Era. The extreme isolation of New Zealand over the last 65 million years must also strongly suggest that there was subsequently little extraneous addition to the cargo of plants and animals which had been its complement when it drifted off on its own.

Therefore, the Fossil Record of the Cretaceous flora of New Zealand also tells us what the contemporary flora of Antarctica and Australia was like. It enables us to ascertain what plants had grown in these other lands, even if they have not been recorded in their own fossil records. Furthermore, the Plant Fossil Record of the Tertiary in New Zealand continues to be a record of families of plants which were already in Gondwana in the Late Cretaceous, and which may be absent from Cretaceous records in New Zealand because of the gaps which are a feature of all fossil records. If there are families in the New Zealand Tertiary which are also present in the Tertiary of Australia, then the implication is that such families had been around in the Late Cretaceous in order for them to have entered New

Reconstructions of a Plesiosaur and a Mosasaur from the Early Creataceous of New Zealand. (Courtesy of Canterbury Museum.)

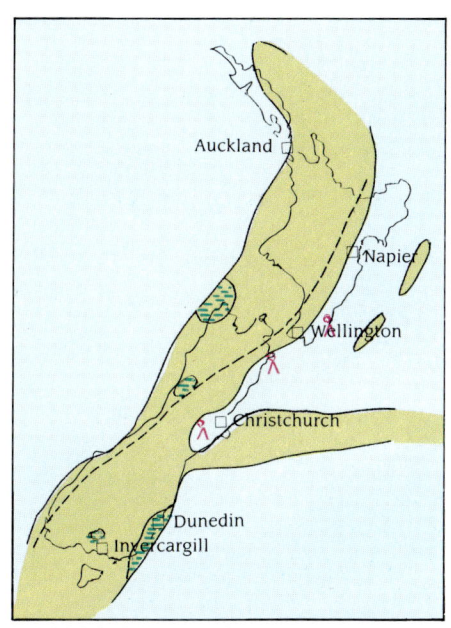

PALAEOGEOGRAPHY OF NEW ZEALAND DURING THE LATE CRETACEOUS EPOCH

NANNOPLANKTON

Coccoliths — microscopic, calcareous plates a few micrometres in diameter — have been important constituents of chalk deposits since Jurassic times. These minute organelles are segments of the outer, calcareous shell of spherical unicellular Green Algae. The Algae are known as Coccolithophores because they bear coccoliths which are amazingly decorative little umbrellas and plates. It is rare to find complete organisms — the ultra-microscopic plates are separated and accumulate in their billions to produce white chalk (limestone) deposits.

SAMI SHAFFIK

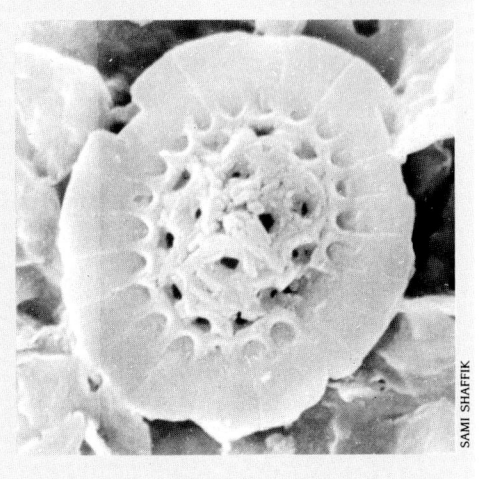

SAMI SHAFFIK

Zealand. This situation is unique: it is possible to know what stage had been reached in the evolution of the Flowering Plants during the Cretaceous by a knowledge of fossil and living flora of New Zealand. And it is surprising to see how many families of Flowering Plants were already developed. Yet it should not be so surprising because bursts of evolution have been a feature of the history of plants and animals, and there were some 40 million years from the first evidence of Angiosperm pollen in the Fossil Record to the time of separation of New Zealand from Gondwana and hence to this stage of development of the Angiosperms.

The Pollen Record of the Late Cretaceous in New Zealand contains an impressive number of families of Flowering Plants, especially when compared with Australia where *Ilex*, *Nothofagus* and Proteaceous-type are the only ones recorded. In New Zealand PROTEACEAE, *Nothofagus* (FAGACEAE), EPACRIDACEAE, CHLORANTHACEAE, HALORAGACEAE *(Gunnera)*, LORANTHACEAE, LILIACEAE, CARYOPHYLLACEAE, WINTERRACEAE, SCROPHULARIACEAE and *Ephedra* are the major pollens identified, and there are many others (mostly not yet identifed) which indicates that this was a very comprehensive assemblage of Angiosperms. The vegetation was rich in Ferns and Tree-ferns including the modern genera *Blechnum*, *Cyathea*, *Dicksonia*, *Gleichenia*, *Hymenophyllum* and *Pteris* (bracken). Conifers were abundant, as they were elsewhere in the Cretaceous, and *Podocarpus*, *Dacrydium*, *Microcachrys*, *Araucaria*, *Agathis* and *Phyllocladus* grew in the forests. The larger Conifer species probably grew as forest emergents, and some of them formed monodominant stands in a manner similar to that of their descendants — the Totara, Rimu and Kauri of modern New Zealand.

The important Broad-leaf macroflora from South Island currently being investigated at Christchurch University will add greatly to knowledge of Cretaceous Flowering Plants in all southern lands. Some of the leaves this macroflora contained resemble Poplar leaves and some have drip-tips like leaves from wet warm-temperate forests.

Vertebrate remains in New Zealand's Cretaceous rocks do not include any Mammals, but Sauropod bones as well as those of marine Turtles, Ichthyosaurs and Plesiosaurs are known. There are also some bones which are believed to be from the foot of a large Running Bird.

The great Running Birds are an example of animals with a Gondwanan ancestor. South America has its Rheas. Africa has its Ostriches. Australia has the Emu and the Cassowary, and formerly had very large prehistoric Birds which only died out in recent times. Madagascar had its Elephant Birds. New Zealand had its Moas within Maori memory, and still has the Kiwi which is descended from the same stock.

New Zealand acquired its ancestral Running Birds from Gondwana before separation, and evolution-in-isolation resulted in distinctly different though closely related big Birds in its fauna. In the absence of Mammals, these Birds assumed the role of grazing animals and had a profound effect on the vegetation.

In the late Cretaceous marine sediments in New Zealand, a prolific Invertebrate fauna similar in composition to faunas of the same age in Australia is found. It includes Ammonites, Belemnites, Bivalves, Gastropods and Brachiopods.

NOTE ON PERIODICITY IN EXTINCTION EVENTS, AND CYCLES IMPOSED ON THE EARTH BY THE DYNAMICS OF THE UNIVERSE

As a result of the controversy surrounding the Terminal Cretaceous Event and the iridium layer, scientists from disciplines as diverse as Astrophysics and Palaeontology have entered the debate about cyclic events which seem to relate to rhythms that are galactic rather than terrestrial. (The use of the word "extraterrestrial" is deliberately avoided here as it has acquired a connotation of visits of flying saucers and intervention in our affairs by "green men" from Mars.)

Earth Science has suffered because of insular, or Earth-bound, vision and perspective. Just as we need to see the local environment as part of the global entity, so we should see the global environment as undoubtedly affected by our planet's relationships to other heavenly bodies and to events which are universal. When we know that the Moon controls a monthly cycle in the seas, and the orbit of the Earth around the Sun creates an annual cycle of changing seasons, there have to be rhythms and cycles which are imposed by the dynamics of the greater system in which Earth is an integral but insignificantly small part.

Statistical analysis of the Fossil Record now seems to show an approximately 26-million-year periodicity in extinction clusters. (These larger-than-background extinctions are pulses when more species went extinct than in the usual extinction/renewal pattern of evolution.) Meteorite impact craters seem to follow a similar pattern, with their ages clustered round 26- to 30-million-year intervals. Matters like magnetic reversals (in which there is a flip in polarity, with North becoming South and vice versa), their frequency and their possible effects on the biota are being debated, as are a host of other unexplored and not explained phenomena. Cosmic "weather patterns" with cyclic rhythms should come as no surprise. The concept only brings the awesome wonder of the Universe home to earthlings.

It may well be that times when the Earth was showered with meteorites and occasionally hit by big asteroids

REG MORRISON

Star trails at Gosse's Bluff. The shattercone from a meteorite impact is seen silhouetted against the night sky at Gosse's Bluff, south of the MacDonnell Ranges in Central Australia. This picture epitomises Earth's interaction with other heavenly bodies. The meteorite crashed into the Earth about 130 million years ago, creating a crater 25 kilometres in diameter. The crater rim has since eroded away. Gosse's Bluff is a massive circular range of mountains enclosing a pound. It represents the centre of impact — the shattercone made from rock which burst to the surface from a great depth at the time of impact. Shockwaves from an impact of this magnitude would have been felt all over the world.

were times when, for cosmic reasons, climatic change has been superimposed on Earth rhythms. Increased volcanism is also apparently cyclic, and there may perhaps be a pattern of periodic acceleration in plate movements, with both phenomena being responses to the beat of a bigger drum — the dynamics of the Cosmos about which we know next to nothing.

RADIOLARIA

Radiolarians have existed since Cambrian times, 570 million years ago, and are still one of the main constituents of marine plankton today. Alive, or as skeletons, they are among the most amazingly beautiful animal forms. They have an intricate skeletal structure, with a central perforated capsule and radiating rods. In the living animal, the central capsule separates a body of inner cytoplasm from the cytoplasm of the outer layer.

Radiolaria, showing the complex and remarkable structure of these beautiful acellular creatures which are an important component of plankton.

JOHN CLEASBY AND GUNTHER BISCHOFF

JOHN CLEASBY AND GUNTHER BISCHOFF

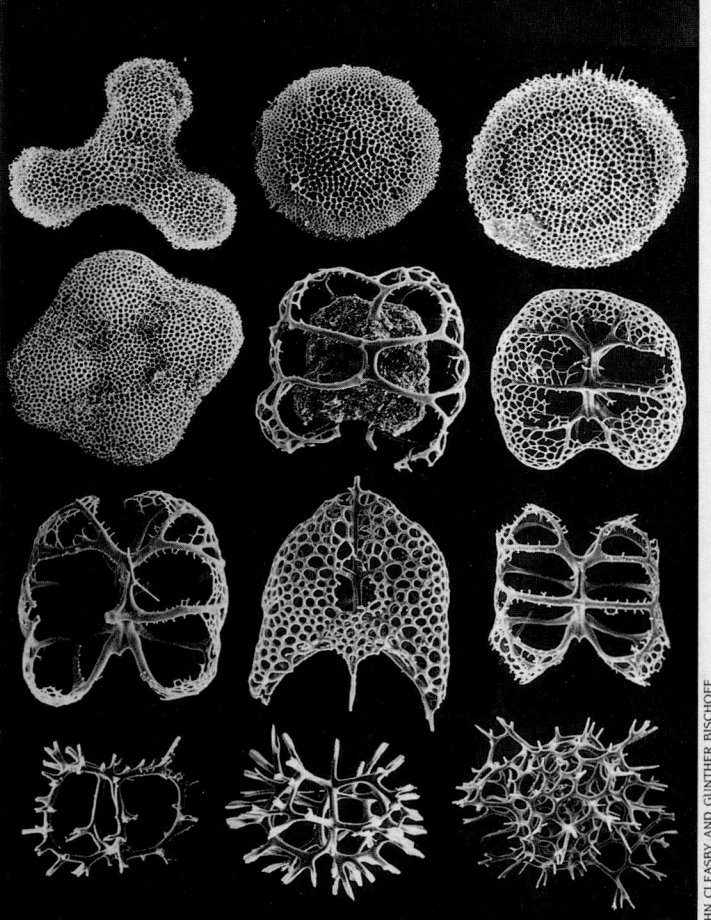

JOHN CLEASBY AND GUNTHER BISCHOFF

THE CAINOZOIC (MOST RECENT) ERA

The Cainozoic Era comprises the last 66.4 million years of Earth history. It has traditionally been divided into two Periods, the Tertiary Period of 64.8 million years' duration and the Quaternary Period of the last 1.6 million years. Now, however, some geoscientists further subdivide the Tertiary into two parts: the Palaeogene, from 66.4 to 23.7 million years ago, and the Neogene, from 23.7 to 1.6 million years ago. (The derivation of the names is from the suffix *gene*, which means "generated by" or "bringing in"; *palaeo* is "the old", *neo* is "the new".) So, effectively, the Cainozoic is sometimes considered as having not two, but three distinct "Periods" — the Palaeogene, the Neogene, and the Quaternary. To minimise confusion, this book maintains the traditional division of the Cainozoic and treats the Palaeogene and Neogene as major divisions of the Tertiary Period.

The Palaeogene comprises the Palaeocene, the Eocene and the Oligocene Epochs, and was a time when the ancient world was changing progressively towards one with more modern features. In the Miocene and Pliocene Epochs of the Neogene, which were to follow, the modern world was finally born out of the ancient one.

The Quaternary Period is divided into the Pleistocene Epoch which encompasses the last ice age, and the Holocene Epoch, consisting of the most recent interglacial in which we are living.

The geological history and the Fossil Record of the Palaeozoic and Mesozoic Eras documented a changing world in which an ever-changing succession of "prehistoric" plants and animals evolved, thrived and became extinct. The arrangements of land and sea on the surface of the globe altered considerably through the long ages, but by late in the Mesozoic the delineation of the margins of modern continents had been accomplished. The Cainozoic Era saw progressive change towards the world of today, and the modernisation of the biota.

Sea level was markedly higher at the start of the Cainozoic Era than it is today — possibly as much as 100 metres higher. Even the short, sharp sea level drop which was a feature of the Terminal Cretaceous Event had resulted in a level about 50 metres higher than that of the present.

The sudden cooling episode at the end of the Cretaceous heralded the deterioration of global climates from the long-established warm to hot conditions which characterised the Mesozoic. In the Cainozoic Era, fluctuating temperature regimes and recurrent glaciations would slide towards the Pleistocene ice age. The abundance of newly formed high mountain ranges, which were a consequence of plate movements, had an effect on continental weather patterns and they were the sites for glaciers during cold spells.

The establishment of a circum-polar current — as Australia moved away from Antarctica, and the Drake Passage opened between South America and Antarctica — was to have an increasingly strong influence on world weather as the Cainozoic progressed. It contributed to the cooling of the South Pole and the establishment of the permanent ice cap. In the case of Australia, the northwards movement of the continent away from high latitudes compounded the effects of the changing world climatic regimes.

The evolution of the Australian environment, in particular, is intimately bound up with changing conditions in Antarctica during the Era. To begin to appreciate this geological past, the role of modern Antarctica in determining present-day global climatic patterns has to be understood.

The Antarctic ice cap maintains a steep temperature gradient from Pole to Equator, producing active ocean and atmospheric circulation. The cold abyssal water (that of the deep-sea bottom) is generated round the ice cap and circulates round the globe, taking about 400 years to travel through the Atlantic to Greenland, and 1500 years through the Pacific. Abyssal water is very cold, about 2° Centigrade, and very salty. It is formed from the melting of sea-ice round the edges of the ice cap. The water sinks to the

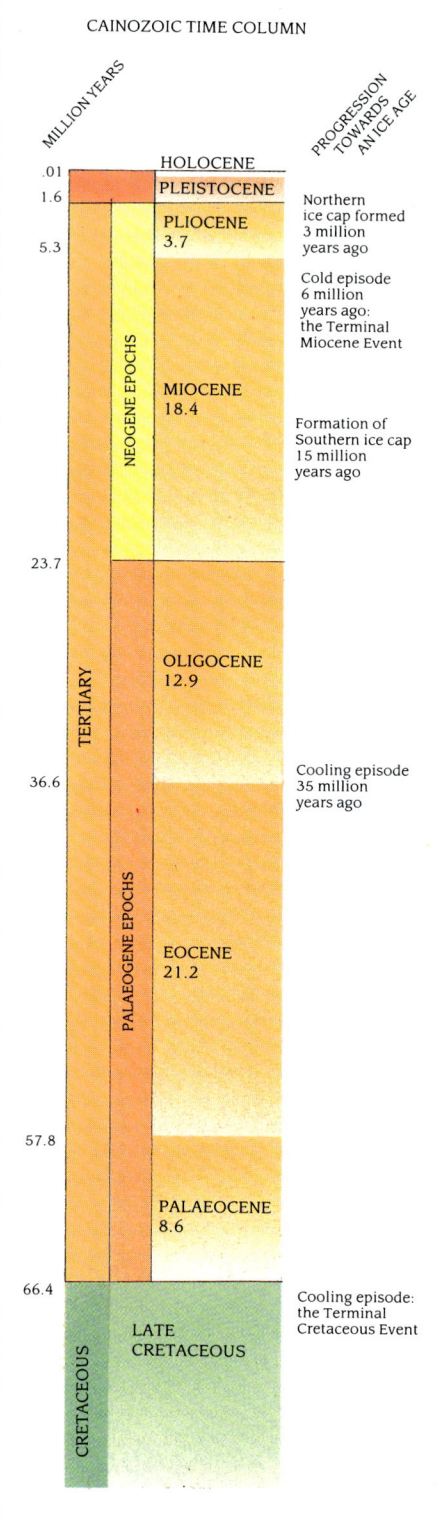

CAINOZOIC TIME COLUMN

MILLION YEARS

PROGRESSION TOWARDS AN ICE AGE

HOLOCENE
.01
1.6 — PLEISTOCENE

PLIOCENE 3.7
5.3

Northern ice cap formed 3 million years ago

Cold episode 6 million years ago: the Terminal Miocene Event

NEOGENE EPOCHS

MIOCENE 18.4

Formation of Southern ice cap 15 million years ago

23.7

OLIGOCENE 12.9

TERTIARY

36.6

Cooling episode 35 million years ago

PALAEOGENE EPOCHS

EOCENE 21.2

57.8

PALAEOCENE 8.6

66.4

Cooling episode: the Terminal Cretaceous Event

CRETACEOUS

LATE CRETACEOUS

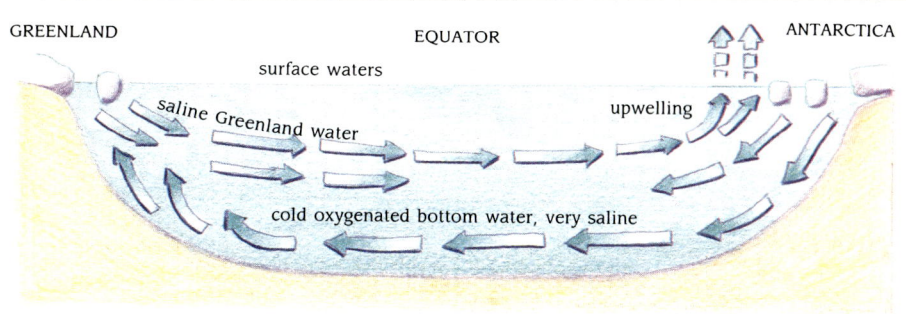

ABYSSAL WATER CIRCULATION
From Antarctica to Greenland via the Atlantic

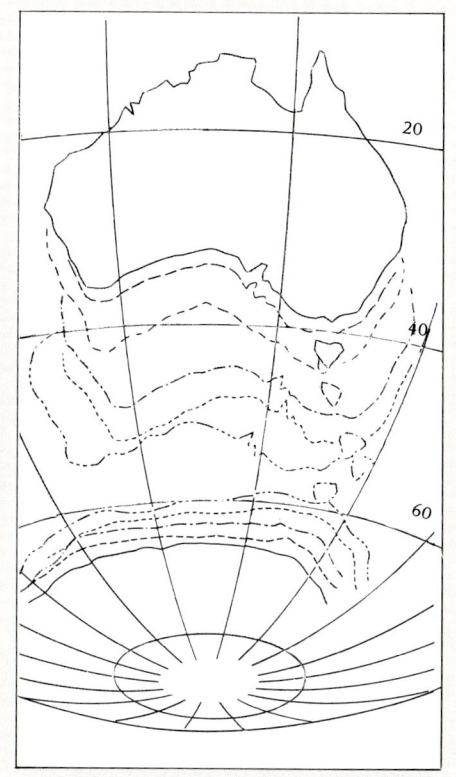

AUSTRALIA'S NORTHWARD MOVEMENT
AWAY FROM ANTARCTICA

———————	Present
— — — — —	Late Miocene
– – – – – –	Early–Mid Miocene
–·–·–·–·–	Early–Mid Oligocene
·············	Mid–Late Eocene
–··–··–··–	Palaeocene–Early Eocene

The very cold abyssal water which results from the melting of sea-ice on the edge of Antarctica is very saline. It sinks and travels northwards as an abyssal current along the bottom of all ocean basins. (Currents at depth in the sea are the result of differences in temperature and salinity which produce density contrasts.) Further from Antarctica's edge, icebergs melt and another water mass results which is not quite as cold and also less saline. This cold water mass travels a shorter distance northwards, at a shallower depth. When the abyssal water reaches Greenland it has warmed sufficiently to form a middle layer on its return journey, and it is joined by cold water of similar salinity which is produced round the northern ice cap. This combined middle layer travels south and upwells against the colder water masses near Antarctica, causing snowfalls on Antarctica.

The edge of the Amery Ice Shelf in Antarctica. The melting of sea ice at the edge of the Antarctic ice cap results in a body of very cold and very salty water that sinks to become part of the abyssal current.

BOB TINGEY

CAINOZOIC OCEAN CIRCULATION

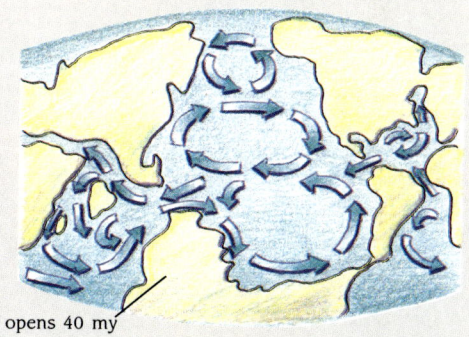

opens 40 my

PALAEOCENE 60 million years ago

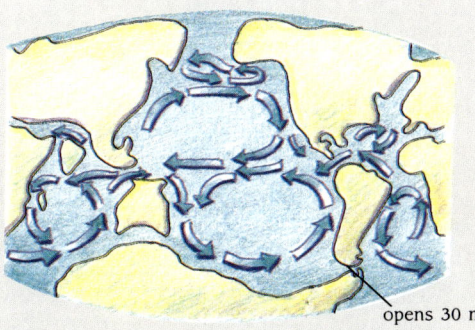

opens 30 my

LATE EOCENE 40 million years ago

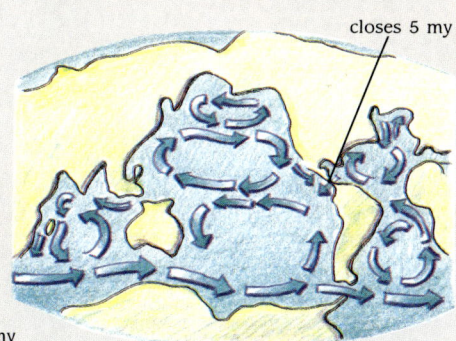

closes 5 my

EARLY MIOCENE 20 million years ago

bottom and forms an abyssal current. The melting of icebergs in a zone round Antarctica produces another body of very cold water which also sinks. But this water is not as cold or as salty as the abyssal water, so it forms a middle layer (between the surface water and the abyssal water) and travels northwards as well, though for a comparatively shorter distance. In the Atlantic, the coldest bottom water reaches North Polar regions off Greenland sufficiently warmed by its long journey to rise and begin its return journey to the South Pole as a middle layer. Upwelling of this middle layer against the colder Antarctic surface layers generates moisture, and the resulting precipitation is in the form of snow which falls on Antarctica and keeps up the supply of ice.

The northwards flow of abyssal water is dependent on the presence of deep channels which connect ocean basins. Two such connections are a trench to the east of the Falkland Islands in the south-western Atlantic, which is vital for its Atlantic passage, and a deep channel near Samoa which controls the flow of abyssal water to the northern Pacific.

The circum-polar current (a current circling the Pole) in surface waters round Antarctica prevents warm equatorial waters from reaching high latitudes, keeping the South Pole cold and maintaining cool and humid conditions in high latitudes. The westerlies, which spin off round the high pressure cell situated over the Antarctic continent, advance in a series of fronts over southern Australia and the southern parts of South America, Africa and New Zealand. Satellite photographs show these weather systems dramatically.

The conditions existing in Antarctica today and the influences they have on the modern world are the result of processes that were evolving throughout the Cainozoic. The separation of Australia and Antarctica was necessary for the formation of the circum-polar current. By the end of the Mesozoic, the rift zone between Australia and Antarctica had been invaded from the west by an eastward spreading arm of the Indian Ocean. It formed a narrow seaway along the

southern margin of Australia, leaving only the Tasmania–South Tasman Rise region as a trailing edge keeping a connection between the two continents. Early in the Tertiary the narrow seaway had become a broad gulf — the early Southern Ocean — and about 45 million years ago the hinge through Tasmania was inundated. It was to remain a shallow area and prevent the development of a strong current until about 38 million years ago. Then the hinge area south of Tasmania was breached at depth and Australia's journey as a drifting island-continent began. Shortly after, about 35 million years ago, the Drake Passage started to open, enabling the progressive development of the current. As Australia proceeded northward, the Southern Ocean widened, the current intensified and the South Pole cooled progressively.

Meanwhile, the separation and movement of other lands, as plates moved and ocean basins widened, altered other circulation patterns. While changes in the Southern Hemisphere were increasingly freeing the development of the circum-polar current, the northward movement of lands resulted in severe restrictions of the equatorial current. Africa moved and pinched the Middle and Near East against Laurasia, closing what was left of the Tethys Sea and isolating the Mediterranean Sea Basin. The closing of the gap between Spain and Morocco closed the western end of the Mediterranean basin. Australia's northward progress eventually deformed the region to its north, blocking most of the flow from the Pacific Ocean into the Indian Ocean. The rising up of the Isthmus Of Panama to link North and South America only 5 million years ago closed yet another gateway for the equatorial current.

All these interactions resulted in the weakening of equatorial flow progressively through the Cainozoic Era. Waters turning north and south from equatorial regions were cooled sooner, ever-cooler water was returned towards the Equator along the western margins of continents, and the consequence of this was the eventual development of the west coast deserts which characterise continents today.

Satellite picture of the weather, from which the chart is made.

The progression towards cooler conditions was not a steady one during the Cainozoic Era. Although the general trend was towards an ice age, there were fluctuations. About 35 million years ago, on the Eocene–Oligocene boundary, there was a pronounced cooling episode. Another in the Mid Miocene, about 15 million years ago, coincided with a global drop in sea level, indicating the formation of the southern ice cap. From then the world was locked into an ice age regime. The next dramatic cooling was at about 6 million years ago. It was accompanied by another sea level drop, indicating a large increase in ice volume. This sea level fall had a profound effect on the Mediterranean Sea, stranding it and causing it to evaporate and refill several times. A brief warming in the Pliocene, between 5.3 and 3 million years ago, was followed by the formation of the northern ice cap at 3 million years ago. The ice age was then fully operational.

Fluctuations between glacial and interglacial stages within the ice age, with increases and decreases in the size of the ice caps, caused significant sea level changes. In glacials more water was held captive in ice so sea levels fell worldwide; in interglacials ice melted and sea levels rose. Many modern landscapes show evidence of these fluctuations in sea level; for example, the sculpturing of coastlines, benches cut in coasts at previous sea levels, and river channels cut deep in order to reach the sea when its level was lower. The evolution of the famous Sydney Harbour, Broken Bay and other "drowned valley systems" which characterise the eastern coast of Australia is related to sea level changes in this ice age.

At times of lowered sea level, the continental shelves of landmasses were exposed and much more land appeared. The Australian land at these times included Tasmania, New Guinea and a wide strip all round the continent. An almost continuous landbridge extended from northern Australia to South-east Asia at the peak of the ice age when the sea level was at its lowest. The continental shelf exposed round New Zealand, particularly round the South Island which was heavily glaciated, was significant because it offered a refuge for vegetation driven off the higher ground by the advancing ice in glacials.

ANTARCTIC WEATHER MAP

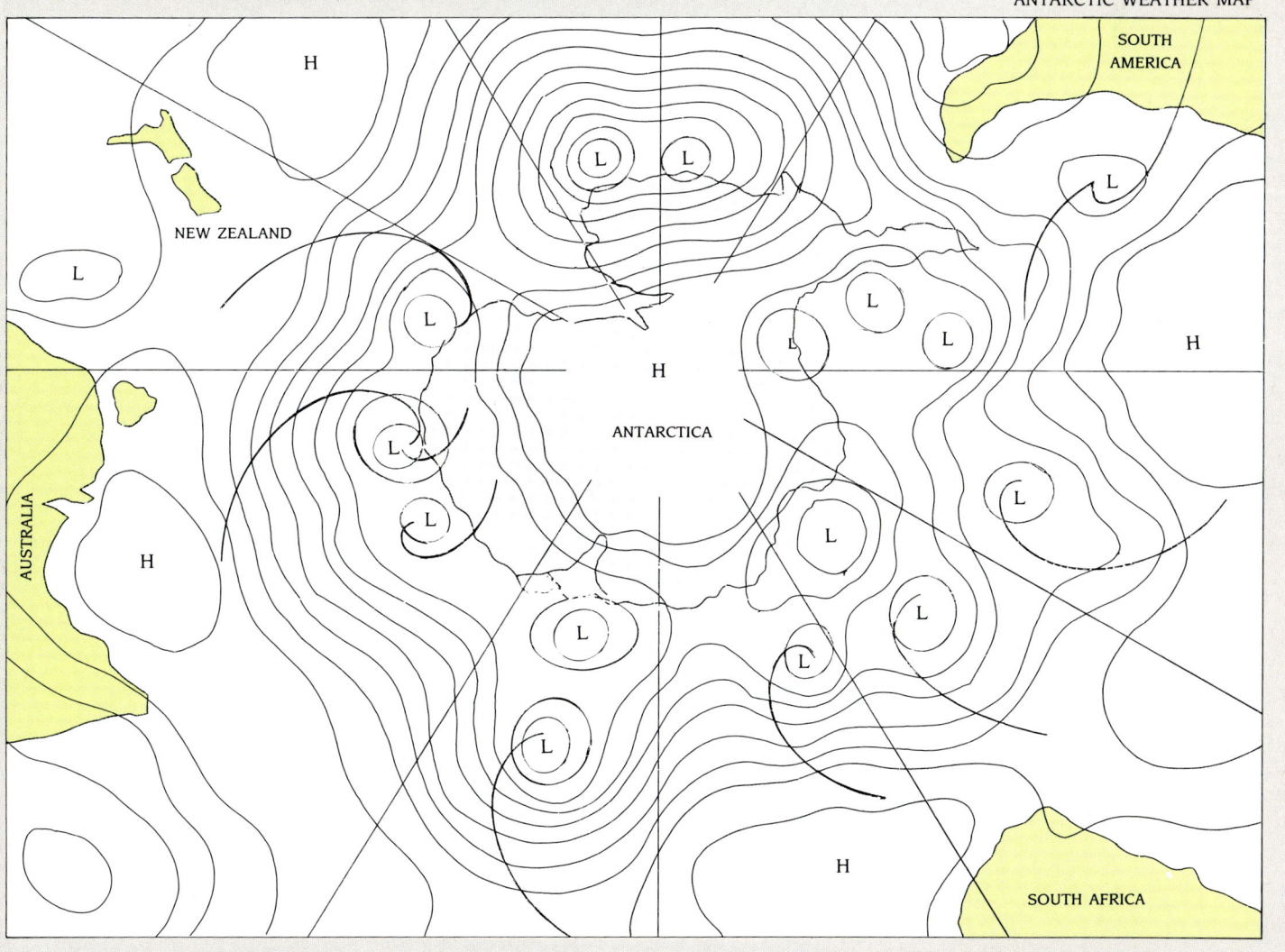

CHAPTER 9

FLOWERS AND FRUITS, MAMMALS AND MILK

THE TERTIARY PERIOD
FROM 66.4 TO 1.6 MILLION YEARS AGO
DURATION: 64.8 MILLION YEARS

The Tertiary Period is characterised by the dominance of Mammals in the Animal Kingdom and of Flowering Plants in the Plant Kingdom.

During the first half of the Period there was a great diversification of Marsupials in Australia. As well, a fauna rich in Reptiles and Birds evolved in the Broad-leaf Rainforest which covered much of the continent. The second half of the Period saw increased climatic change as ice age regimes became a feature of world climates. Aridity was the main factor affecting the Australian environment as the Period progressed, resulting in the establishment of the arid Centre, the spread of scrub and grasslands, the contraction of Rainforest and the evolution of schlerophyll vegetation.

New Zealand was a lowland plain, steadily eroding and being invaded by the sea until the latter part of the Tertiary when its re-emergence as rising land occurred. Its Gondwanan complement of plants and animals evolved in isolation and over time many species became extinct due to the environmental stresses which tectonics and climatic deterioration towards an ice age imposed on the biota.

The Wet Tropical Rainforests of Queensland are confined to small refuges where the environment has remained suitable for them while the rest of Australia has been drying out. These remnant forests represent the Gondwanan forest of 60, or even 80, million years ago.

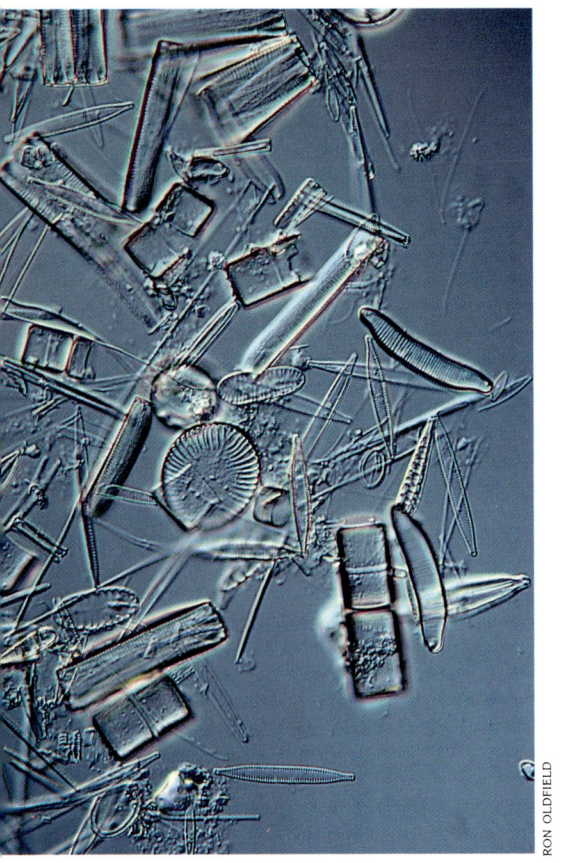

RON OLDFIELD

Mixed Diatoms.

A leaf of a member of the family PROTEACEAE, from Penrose, New South Wales. (Specimen AMF. Magn.X 2.0)

The Tertiary Period was named in 1759, but since that time so much information has become available about the geology and palaeontology of the last 65 million years of Earth's history that a logical separation of the Period into two parts — the Palaeogene and the Neogene — is becoming generally accepted.

The wealth of information results from the abundance of rocks of Tertiary age. Erosion has had less time to remove the evidence. In addition, modern technology applied to the study of deep-sea sediments (oxygen isotope studies, in particular) has supplied a great deal of information about ancient climates and conditions.

The Palaeogene encompasses the first two-thirds of the Tertiary, from 66.4 to 23.7 million years ago, and comprises the Palaeocene, Eocene and Oligocene Epochs. The Neogene covers the last third, from 23.7 to 1.6 million years ago, and comprises the Miocene and Pliocene Epochs.

During the Palaeogene Epochs the changing arrangements of land and sea together with the cooling of the South Pole progressively altered world weather patterns. For most of this time climates were warm and humid. The sudden cooling associated with the Terminal Cretaceous Event had been a warning of changes to come, even though it did not persist. Though the hot and bountiful climes of the Cretaceous were never to return, conditions were still ideal for the spread of temperate Rainforest in South America, Antarctica, Australia and New Zealand. After the slow start in the Mid Cretaceous, great diversification of Flowering Plants and their rapid dispersion resulted in widespread development of Broad-leaf forest, which persisted till the Mid Miocene.

At the end of the Eocene there was a sudden marked cooling worldwide, and at various times there may have been other temperature fluctuations. It is not possible to determine how many of these colder/warmer alternations occurred as only major changes can be detected. (It has been shown that in the Pleistocene ice age cooling events with glaciation came on suddenly over thousands rather than millions of years, and that warming phases were also abrupt.)

Tectonic events in the remote Northern Hemisphere were to affect the South Pole by the middle of the Miocene Epoch, with the separation of Greenland and Norway and the sinking of the Faeroe–Iceland ridge creating a new circulation pattern. The North Atlantic water now travelled south, welling up near Antarctica and resulting in greatly increased snowfalls which accelerated the formation of the ice cap. At the same time the Australian Plate was colliding with the Pacific Plate, and New Guinea was starting to rise above the sea and would soon begin to form mountainous terrain. The lowered sea levels caused by the increased size of the ice cap resulted in the creation of a great deal of land to the north of the present Australian continent.

Late in the Miocene, the climate in high latitudes was colder than it had been at any time since the Permian ice age. Between 6 and 7 million years ago there was a sudden and dramatic cooling. The Ross Ice Shelf in Antarctica expanded to extend up to 200 kilometres further north than it does today, and sea levels fell dramatically. This sudden cooling, the "Terminal Miocene Event", had a number of significant consequences.

The Mediterranean Basin was stranded by the fall in sea level and its waters evaporated to form deep salt deposits. Evidence shows that there were several drying up and refilling episodes corresponding to fluctuations in the size of ice caps. So much salt was taken out of circulation (and still lies in enormously deep layers under the Mediterranean Sea today) that the salinity of world oceans dropped by a staggering 6 per cent. It is fascinating to try to visualise the dry Mediterranean Basin when it was a desert with

salt lakes, islands of vegetation, shimmering white salt plains and endless marshes. The scale of such an event in terms of the volume of water which evaporated, the depth of salt which was precipitated, and the changes that the process must have made on adjacent lands, is almost incomprehensible. There must have been spectacular waterfalls across the Straits Of Gibraltar each time it refilled.

At about the time the Mediterranean Sea was drying up, parts of now arid North Africa were green and fertile, with forests growing on the Atlas Mountains. However, further south the situation was different, with aridification of the west coast region and the formation of the Namib Desert resulting from the upwelling of the cold Benguela Current along the West African coast. Elsewhere in Africa the sudden cooling and drying altered the vegetation. Forest was replaced by grasslands as parts of the land became seasonally arid. The forests continued to retreat over the next few million years and the ancestral humanoid Apes came down out of the trees and started to walk upright on the savannahs. Thus the evolution of *Homo sapiens* started and the problems of Earth's survival as a living biosphere were about to begin.

AUSTRALIA IN THE TERTIARY PERIOD

Tectonically the Australian continent was relatively stable during the Tertiary. Uplift of the eastern highlands probably began on a small scale in the Eocene, peaked in the Miocene and has continued gradually and episodically ever since. Subsidence of the Murray–Artesian depression and upwarping of the western half of the continent proceeded gradually, and there was some faulting and block movement in marginal basins.

Tertiary volcanic activity occurred in a broad but discontinuous belt down the eastern highlands, round into western Victoria and South Australia, and into Tasmania. At different times during the Period a great deal of igneous rock was produced in great basalt flows, sills and dykes, and also in localised volcanic eruptions. Over 30 centres of eruption in Tasmania and an equally large number in Queensland tell of an active period of volcanism. In New South Wales eruptions seem to have been of a fissure type as centres are mainly hard to pinpoint. However, obvious centres were at Mt Canobolas and in the Liverpool Ranges and the Barrington Tops. Many features of modern landscapes are products of the Tertiary volcanism: the Warrumbungle Range and Nandewar Range in New South Wales, the Glasshouse Mountains in Queensland and the newer Victorian basalt plains are a few examples. The fertile soils produced by weathering of volcanic rocks are of great importance in a land such as Australia where ancient soils deficient in many elements tend to be the norm.

Terrestrial deposits of Tertiary age are of two main types: those of relatively small lakes and rivers of the eastern highlands and coastal plain; and those of the great internal drainage systems which covered parts of the old Drummond, Georgina, Amadeus and Great Artesian basins, where rivers spread great sandsheets. Rivers in the Pilbara and Carnarvon Basin region of Western Australia were also spreading sheets of sand. The great Tertiary internal drainage system had several centres. Only the Lake Eyre Basin inward drainage system has retained a degree of permanence.

Eocene to Miocene deposition in the great internal drainage system resulted in the creation of the Etadunna Formation of Late Oligocene age in which an important assemblage of Vertebrate fossils has been found. (Few fossiliferous deposits of Early to Mid Tertiary age are known outside the

The Koala has been around for a long time. A fossil of a Koala that lived about 25 million years ago has been found in Central Australia, near Lake Eyre. This Late Oligocene species was smaller than the modern Koala. When it was alive the Lake Eyre region was a forested land of lakes and streams, unlike the dry desert of today.

JIM FRAZIER

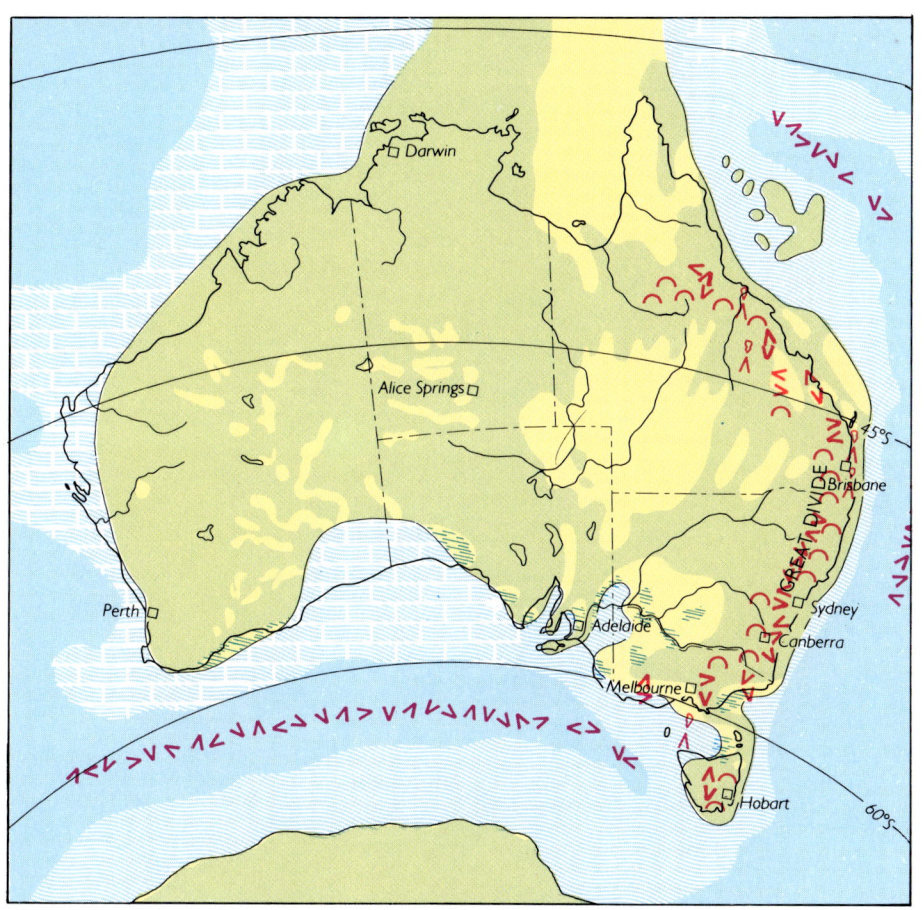

PALAEOGEOGRAPHY OF THE EARLY TERTIARY
60 million years ago

BRUNO JEAN GRASSWILL

Pelicans — with Flamingoes, fresh-water Dolphins, Lungfish and many Marsupials — lived in the Lake Eyre region of Australia 25 million years ago when the area was well-watered and supported forest. Finding the bones of such unlikely animals today in the dead heart of the country comes as a surprise and emphasises how much environments have changed through time.

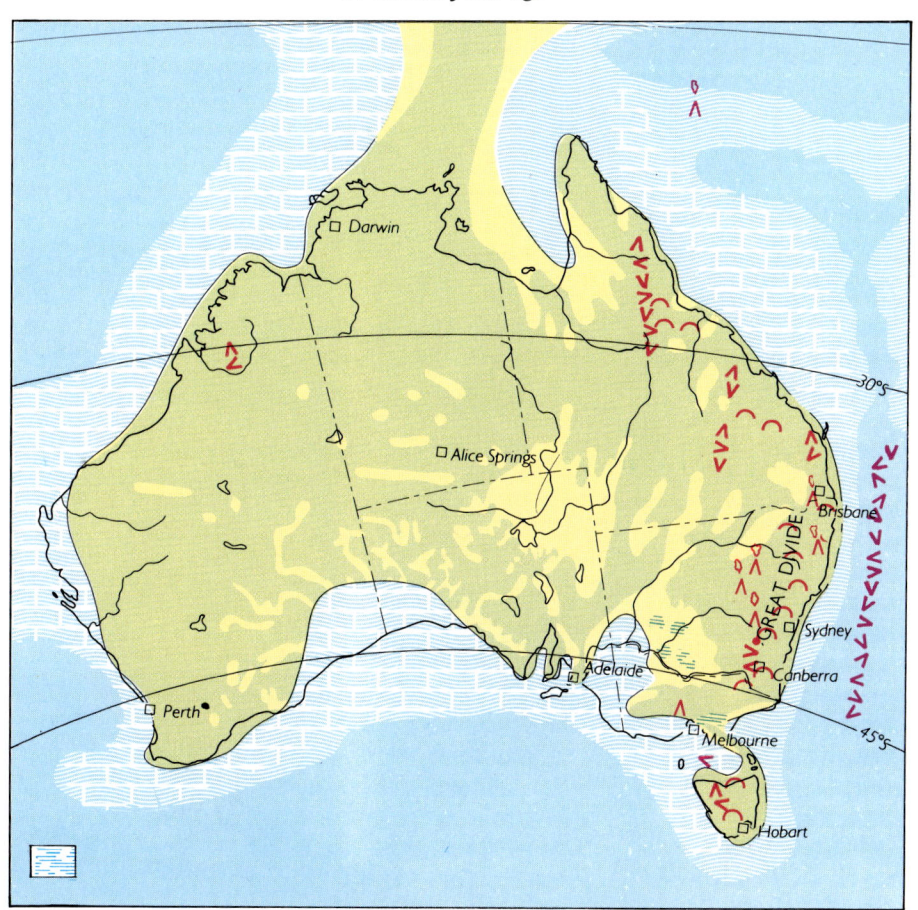

PALAEOGEOGRAPHY OF THE LATE TERTIARY
20 million years ago

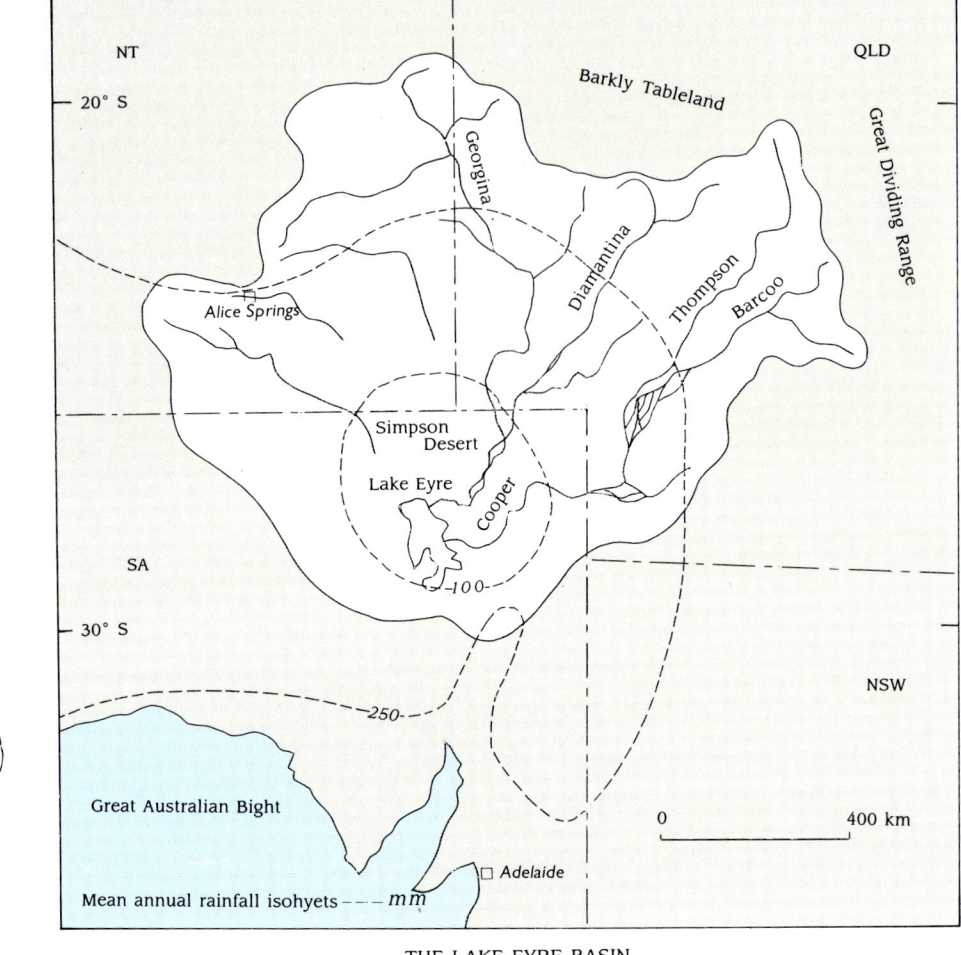

THE LAKE EYRE BASIN
A Great Inward Draining System

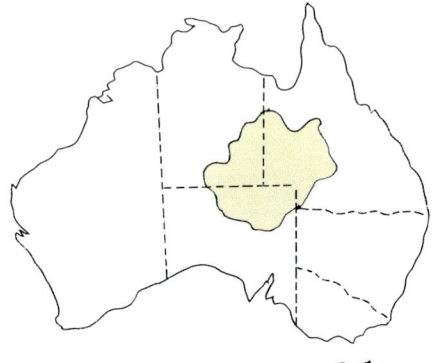

SITUATION OF LAKE EYRE BASIN

Lake Eyre Basin.) In the Late Tertiary erosion was very active in the Centre, resulting in stream channel deposits of which the Pliocene Mampuwordu Sands and the Tirari Formation, with their valuable mammalian fauna, are examples.

Marine Tertiary rocks are known from the Murray and Eucla basins in the south, the Carnarvon Basin in Western Australia, and all around the continental shelf, notably on the Great Barrier Reef platform off the Queensland coast. The Miocene limestones of the Eucla Basin (now the Nullarbor Plain) form shallow soils which have acted as a barrier to the spread of plants from east to west, or vice versa, across the region. The marine Tertiary faunas comprise Brachiopods, Bryozoans, Echinoids, Polyzoans, Gastropods and Bivalves. Microscopic Foraminifera are abundant in plankton and are useful in correlating zones .

Much of the modern Australian landscape consists of land surfaces which came into being during the Tertiary Period and are now in various stages of erosion and dissection. Commonly two, and sometimes more, highly planated surfaces are visible. In semi-arid regions the older land surface is present as outlying mesas which rise above plains of the younger surface and merge laterally into extensive plateaux. The Tertiary surfaces in the more humid eastern highlands were uplifted during the Kosciusko Orogeny, which occurred late in the Period, and the old land surfaces form the tablelands.

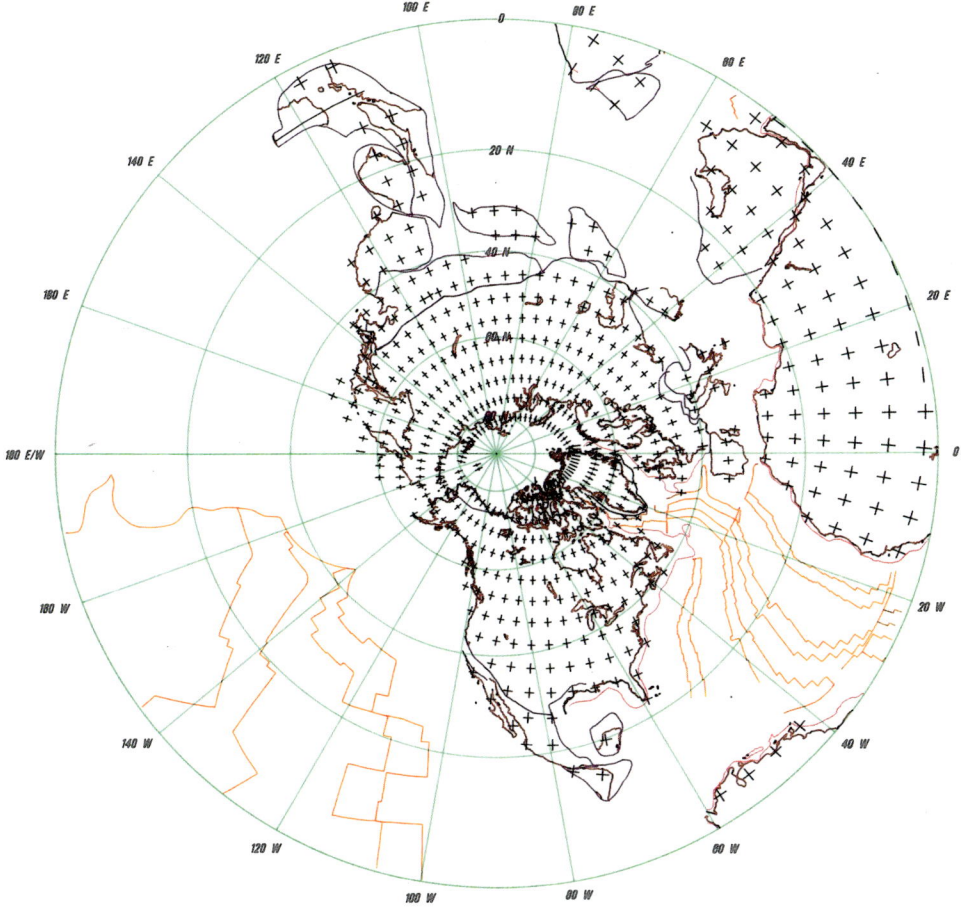

NORTH POLAR PROJECTION

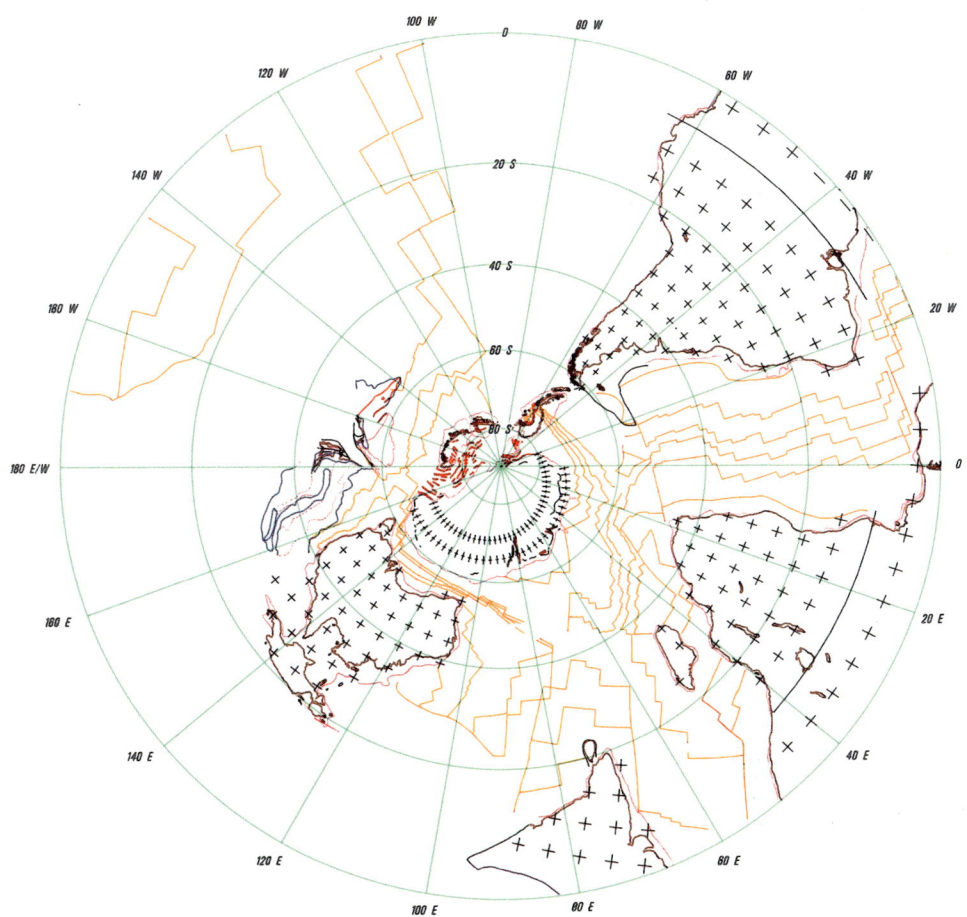

SOUTH POLAR PROJECTION

*Butterflies evolved in tandem with
Flowering Plants during the Tertiary Period.
It was the evolution of efficient pollination
mechanisms that made the Angiosperms so
successful and led to rapid diversification.
Special relationships were established
between Insect species and specific plants,
and also between Birds and Mammals and
some flowers.*

JIM FRAZIER

*POSITION OF LANDMASSES
IN RELATION TO THE
NORTH AND SOUTH POLES
IN THE EARLY TERTIARY,
60 million years ago*

Silicified Banksia stems seen in cross-section. From Springsure, Queensland. (Specimen AMF. Magn.X 1.3)

TERTIARY PLANT MACROFOSSILS

Leaves of Rainforest trees (Angiosperms, Dicotyledons) from the Eocene at Elsmore, northern New South Wales. (Specimen AMF. Magn.X 3.0)

THE PALAEOGENE

FROM 66.4 TO 23.7 MILLION YEARS AGO
DURATION: 42.7 MILLION YEARS

Halfway through the Palaeogene, Australia and Antarctica separated and the island-continent started moving northwards. Rainforest covered much of the land throughout the Palaeocene, Eocene and Oligocene Epochs which comprise the Palaeogene. The composition of the Rainforest changed to match fluctuations in climate which resulted from Australia's northwards progression and from a global trend towards polar cooling. The Marsupial fauna evolved from Gondwanan ancestors, establishing the basis for Australia's unique modern fauna. The Marsupials were able to diversify and fill all niches in the absence of competition from Placental Mammals, which, for some reason or other, had not entered Australia while it was connected to other lands.

THE PALAEOCENE EPOCH

FROM 66.4 TO 57.8 MILLION YEARS AGO
DURATION: 8.6 MILLION YEARS

During the Palaeocene Epoch Australia was still attached to Antarctica but its eastern and western margins had been freed. The climate was warm and wet,· and Rainforest covered much of the land. In the Centre the Rainforest had a high proportion of members of the family PROTEACEAE with more than 20 identifiably different pollen types. (Most of these early forms would later become extinct and would be replaced by pollen of living genera.) *Banksia*-type proteaceous pollen is already identifiable from the Palaeocene. *Nothofagus*, the southern Beech which had been among the earliest Angiosperms to appear in the Pollen Record in the Cretaceous Period, was present in the Palaeocene but not abundant. Fern spores were plentiful, but there was a gradual decline in the amount of Conifer pollen of the Podocarp, Celery-top and Araucarian types which had been so dominant in the Cretaceous. The Flowering Plants had taken over the forests. Pollen of members of many families are recognised, including WINTERACEAE, EUPHORBIACEAE, CASURINACEAE, MYRTACEAE, SANTALACEAE, SAPINDACEAE and RESTIONACEAE. The LAURACEAE, one of the earliest families, has pollen which does not often survive fossilisation. It serves to remind us that absence from the Pollen Record does not necessarily imply absence from the continent.

THE EOCENE EPOCH

FROM 57.8 TO 36.6 MILLION YEARS AGO
DURATION: 21.2 MILLION YEARS

The early Southern Ocean between Australia and Antarctica was widening throughout the Eocene, and by 45 million years ago Australia's trailing edge connection to Antarctica was flooded. It was to remain as a shallow shoaling area until the end of the Epoch, when it was breached at depth and became a free-flowing seaway — and a forerunner of the circum-polar current. Carbonate oozes were deposited on the seafloor of the widening Southern Ocean, testifying to the continued warmth of the seas. The Eucla Basin (which was to become the Nullarbor Plain) was flooded and its limestones are the product of carbonate oozes deposited there.

In the Early Eocene the surface water temperature on the Campbell Plateau adjacent to Antarctica was approximately 20°. But global circulation

Early Miocene organ-pipe basalt near Great Lake in Tasmania.

patterns gradually changed during the Epoch in response to the movement of lands and the widening of ocean basins, and the end of the Eocene saw a marked cooling episode.

At about 50 million years ago, in the Early Eocene, India collided with Asia. Thereafter, the Indian Plate fused with the Australian Plate to become the Australia-India Plate which is one of the major plates of the Earth's crust today.

Australia had a rich and varied flora in the Eocene. Early Eocene pollen assemblages from the Gippsland and Otway basins in Victoria, Nerriga in New South Wales, West Tasmania and the Perth Basin in Western Australia contain elements of a flora comprising: MYRTACEAE, SAPINDACEAE, MALVACEAE, BOMBACACEAE (Baobab family), *Beauprea* (PROTEACEAE), *Anacolosa* (OLACACEAE), *Nipa* (the Mangrove Palm), *Nothofagus* (in low proportions as in the Palaeocene) and Conifer pollen. The presence of *Nipa* Palm pollen is interesting, indicating that both Australia and New Zealand where it also occurs received ancestral Palms from Gondwana. Palms were among the early Flowering Plants and spread to form a Palm Province in low latitudes. Some also spread into high latitudes while the polar regions were warm.

In the Mid Eocene some pollen samples show a great increase in *Nothofagus brassii* type and less diversity of associated forms. Such sampling is interpreted as showing a climatic change. However, the number of samples which the microfossil and macrofossil assemblages so far provide for the 20 million years of the Epoch is proportionately so small that no coherent picture emerges. They represent isolated instances separated by unknown lengths of time. This situation is always a major problem in interpreting information from the distant past.

A macrofossil assemblage from Maslin Bay in South Australia has large numbers of leaves similar to those which form the leaf litter in modern Queensland wet forests, even down to the presence of large numbers of associated Fungi which attest to high humidity.

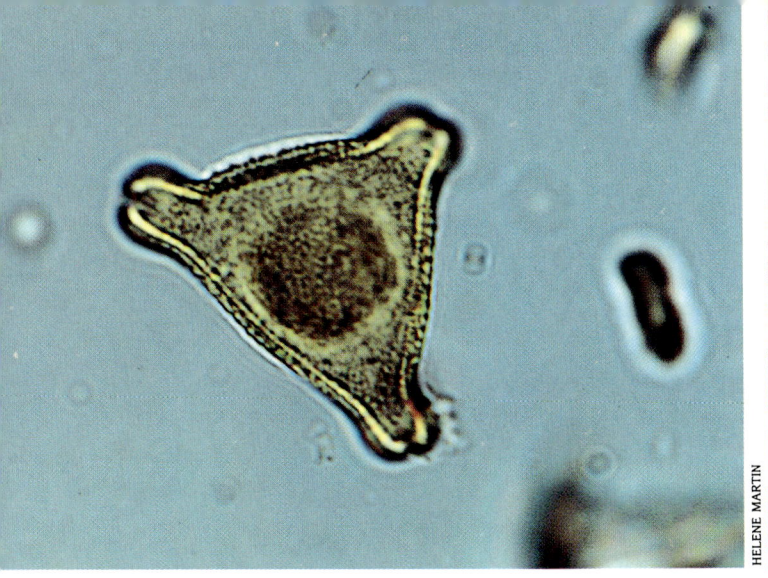
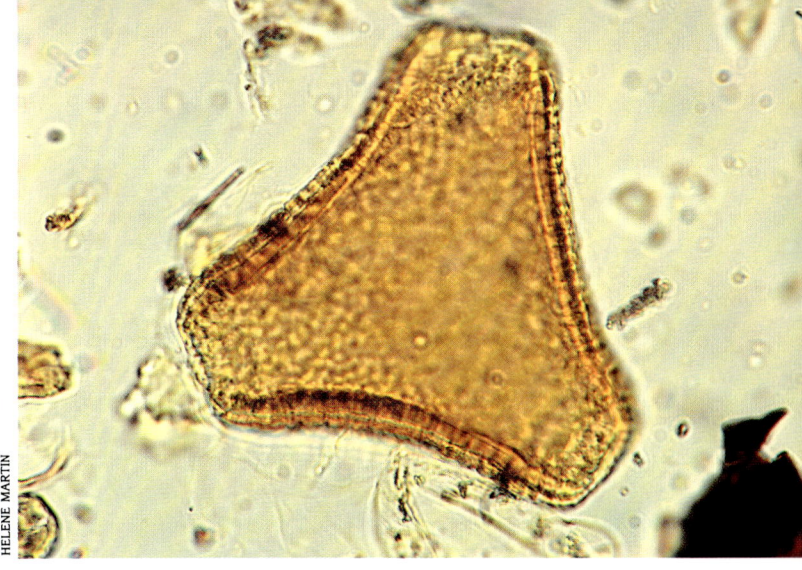

HELENE MARTIN

EOCENE POLLEN FROM EXTINCT PROTEACEAE

Proteacidites pachypolus.

Proteacidites rectomarginatus.

The Vegetable Creek Fossil Flora of northern New South Wales is a rich and diverse assemblage of mainly Dicotyledon leaves with some Conifers. It includes specimens of *Agathis* (Kauri Pine), *Araucaria* (Hoop and Bunya Pines) Podocarps and Celery-top Pines, and is probably Mid Eocene in age. This flora was described by Baron von Ettinghausen in 1880, and the identifications he made were based on leaf form only. Classification of plants on leaf characters, particularly when the fossils are only impressions and no plant tissue remains for close investigation, is a difficult task. But, now that there are pollen samples from rocks of the same age as those in which the Vegetable Creek fossils are preserved, it appears that von Ettinghausen's identifications were surprisingly accurate, at least to family level.

If we had been able to walk through the Eocene forest at Vegetable Creek (in what is now northern New South Wales near the Queensland border) we would have felt at home. We would have recognised many of the trees and forest occupants. Figs (MORACEAE) with Parrots eating their fruits, large Running Birds like Cassowaries in the underbrush, and perhaps Crow-ancestor Birds flying about. We would have heard Cicadas in the canopy where Kurrajong (*Brachychiton*), Sassafras, Laurels (*Cinnamomum* etc), *Sloanea* (ELAEOCARPACEAE), *Callicoma* (CUNONIACEAE), MYRTACEAE (similar to Water Gums and Lilli Pilli) and many others which we know today formed a Closed Forest community. A few Conifers would have been interspersed in the forest, along with vines growing as lianas. Tree-ferns would have been growing in the gullies. And there would have been early Platypuses in the streams, and Possum relatives and other small Marsupials hopping on the forest floor or climbing in the trees above.

In the Late Eocene, for which there is little evidence on the flora, some adaptation may have been required to suit the cooler conditions which prevailed for a while. The surface water temperatures on the Campbell Plateau are known to have fallen to about 10°C and there may have been glaciers on high ground in Antarctica. Pollen samples indicate that southern Western Australia still had Closed Forests. Samples from the Northern Territory contain an abundance of RESTIONACEAE, CYPERACEAE (Sedges) and SPARGANIACEAE (Burr-rushes), indicating that the marshes and water margins were by this time colonised by these specialised plants which had taken over the niche formerly occupied by Horsetails in "prehistoric" vegetation.

The terrestrial Vertebrate Fossil Record for the Eocene in Australia is meagre. Giant Penguins have been found in the Late Eocene Blanche Point Formation at Whitton Bluff near Port Noarlunga in South Australia. These Birds, standing to a height of more than a metre, had long thin necks and their flippers were large when compared with living (or later fossil) species, being closer to a flighted ancestor. The same Giant Penguins have been found in Late Eocene sediments in New Zealand and in Seymour Island in the Antarctic.

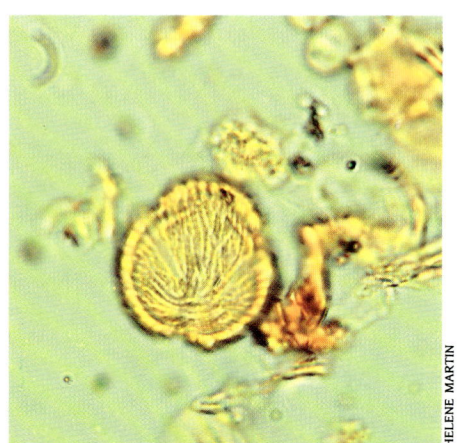

Pollen of ANACARDIACEAE.

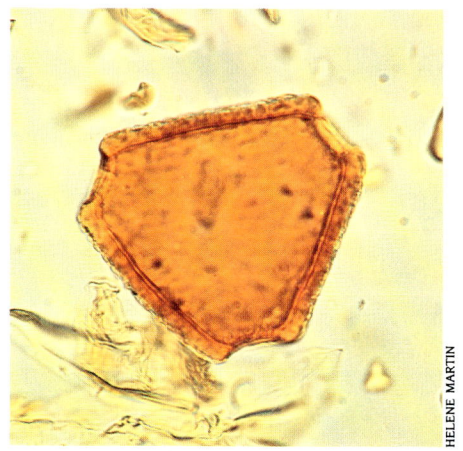

Triorites magnificus, Late Eocene pollen of uncertain affinities.

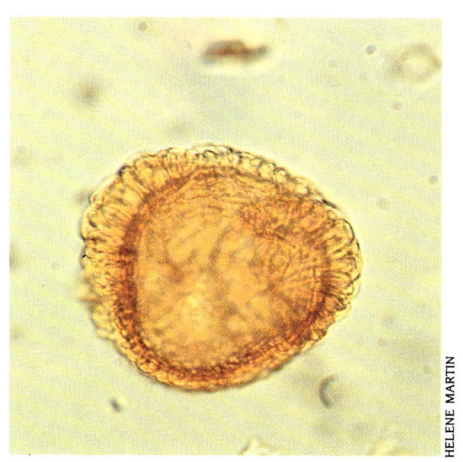

Dacrydium pollen of a type similar to Rimu (Dacrydium cupressinum) of modern New Zealand.

THE OLIGOCENE EPOCH

FROM 36.6 TO 23.7 MILLION YEARS AGO
DURATION: 12.9 MILLION YEARS

At the start of the Oligocene, the surface water temperature on the Campbell Plateau adjacent to Antarctica is known to have been falling. The West Antarctic ice sheet may have been forming, and a pronounced sea level fall in the Early Oligocene relates to its expansion. The gradient in temperatures from Pole to Equator increased as the result of the significant cooling of high latitudes, and water temperatures dropped to 6° or 7°C.

The opening of the Drake Passage between South America and Antarctica in the Early Oligocene (about 30 million years ago) allowed the circumpolar current to develop fully, isolating Antarctica and causing further cooling. As a consequence of the high latitude cooling there would probably have been a northward movement of rain-bearing westerly wind systems. These winds would have affected those considerable parts of Australia that were then lying between the Pole and 40° South, making them well-watered. The north and north-west of the continent may have been considerably more arid, already acclimatising some plants and animals to increased seasonality.

Vegetation persisted in coastal regions of the Antarctic continent until the Late Oligocene. In the Ross Sea area forests dominated by Nothofagus fusca, and with PROTEACEAE, MYRTACEAE, and PODOCARRACEAE, coexisted with the earliest known glaciers in the area. A similar situation occurs in Chile today where the same assemblage is found at the edge of glaciers in the Andes. The PROTEACEAE there includes Embothrium, the closest relative of our northern Australian Oreocallis — a clear example of living plants showing their Gondwanan connections.

In Australia the plant Microfossil Record of the Oligocene is scanty and comes mainly from the south-eastern sector, where it shows a decline in diversity which reflects the colder environment of the time. Nothofagus dominates, with PROTEACEAE, MYRTACEAE and other families. The pollen in the MYRTACEAE is of seven types, including one which resembles Eucalyptus pollen, but it is not possible to assign any of the forms with certainty to any living genera. No plant macrofossils of undoubted Oligocene age are recorded.

Widespread deposition of coals and lignite in south-eastern Australia indicates high water-tables in extensive swamps. Some of the coals contain identifiable pollen, and that of the Conifer Dacrydium is abundant. Today Dacrydium grows in moist, cool-temperate regions.

In the central Australian inward drainage system which was active at this time, the sediments laid down to form the Etadunna Formation documented the Vertebrate population of the times. (This formation was originally considered to be Mid Miocene in age, and has only recently been assigned to the Oligocene.) Lungfish (Ceratodus), other Fish, freshwater Dolphins, Crocodiles, Turtles, Flamingoes, Platypus, Possums, Koalas, Cuscus-like animals, Diprotodontids and early Kangaroos comprise this fauna. Their presence shows that this region, today the driest of deserts, was then a cool-temperate, green and pleasant place with its lakes, rivers and forest supporting a thriving and increasingly modern-type animal population.

Until the Etadunna Formation was reassessed as Oligocene, a Late Oligocene locality on the north shore of the Derwent River in Tasmania had been the only recognised terrestrial Mammal fauna site of that age, however it contained only a few Marsupial bones.

THE NEOGENE

FROM 23.7 TO 1.6 MILLION YEARS AGO
DURATION: 22.1 MILLION YEARS

The Neogene comprises the Miocene and the Pliocene Epochs. The first third of the Neogene was warm to hot, reversing the trend towards the ice age. In the Mid Miocene, 15 million years ago, the permanent southern ice cap formed, and thereafter the cooling trend again became evident. By the end of the Neogene the world was firmly locked into an ice age. Drying out of the Australian continent accelerated as the world cooled, and the vegetation changed. Rainforests contracted, the Centre became desert and the fauna progresssively adapted to the changing environments.

THE MIOCENE EPOCH

FROM 23.7 TO 5.3 MILLION YEARS AGO
DURATION: 18.4 MILLION YEARS

During the first half of the Miocene Epoch surface water temperatures were warmer globally than they had been in the Oligocene. This interval represents the only lengthy reversal in the general cooling trend through the Cainozoic Era. Surface waters in the Tasman Sea were a few degrees warmer than they are now, and the shallow shelf waters on the southern margin of the continent were very warm — between 15° and 20°C, equivalent to subtropical waters today. Circulation was probably fairly sluggish, consistent with times when there is little gradation in temperatures. Sea levels were rising during this phase, mainly due to ice melt but possibly contributed to by collisional tectonics occurring in Asia and in the western Pacific.

Transgressions of shallow seas into the Eucla, Murray, Otway and Gippsland basins in which carbonate oozes were abundant resulted in the formation of extensive limestones in these regions. The southern part of the continent remained well-watered during the Early Miocene.

The fruit of a Dicotyledon. A latex mould made from the impression in the rock is seen next to the specimen. (Specimen AMF. Magn.X 2.6)

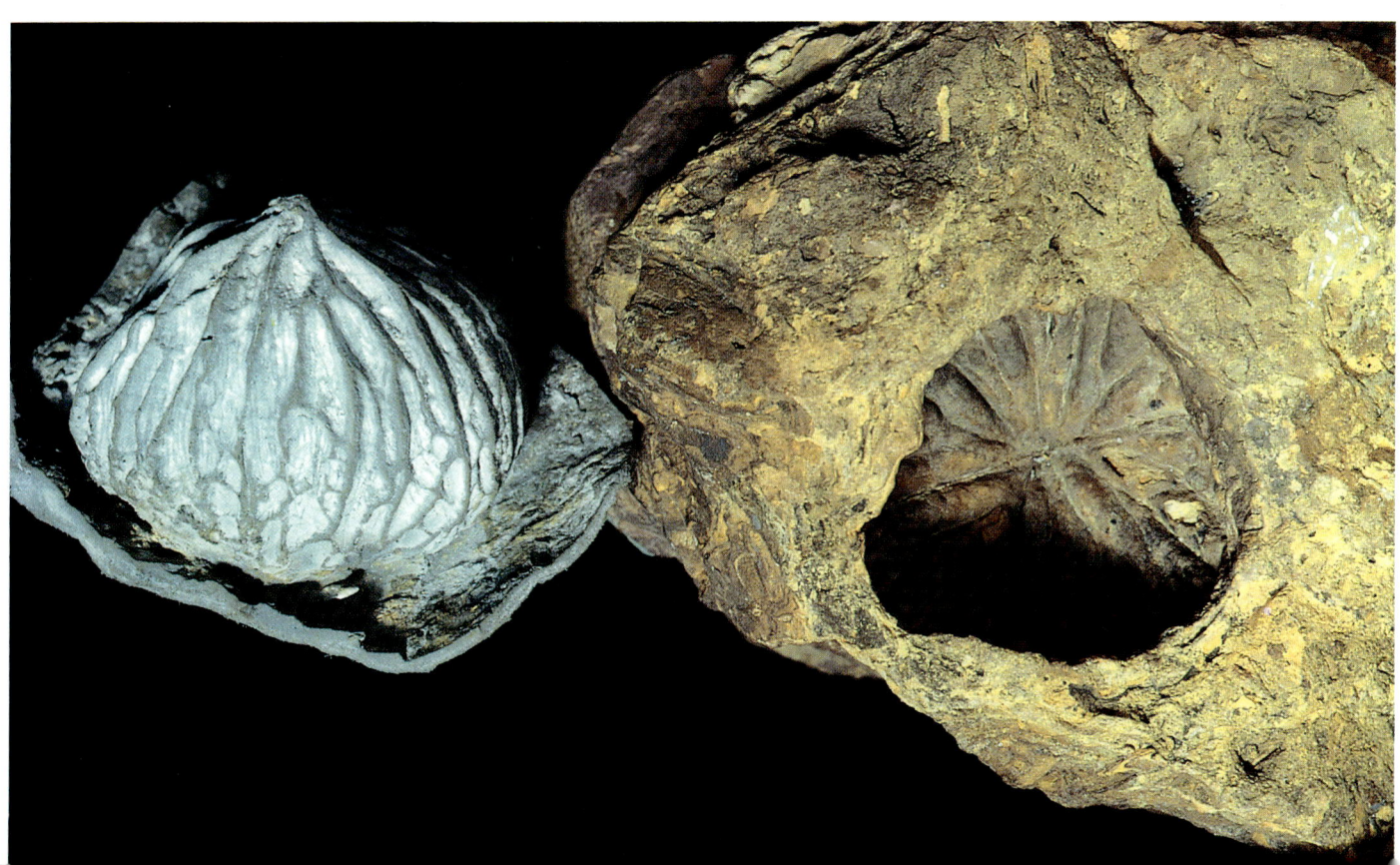

With the development of grasslands the Marsupial ''Antelope-equivalent'', the large Kangaroos with their easy hopping stride, evolved to take advantage of the wide open plains.

JIM FRAZIER

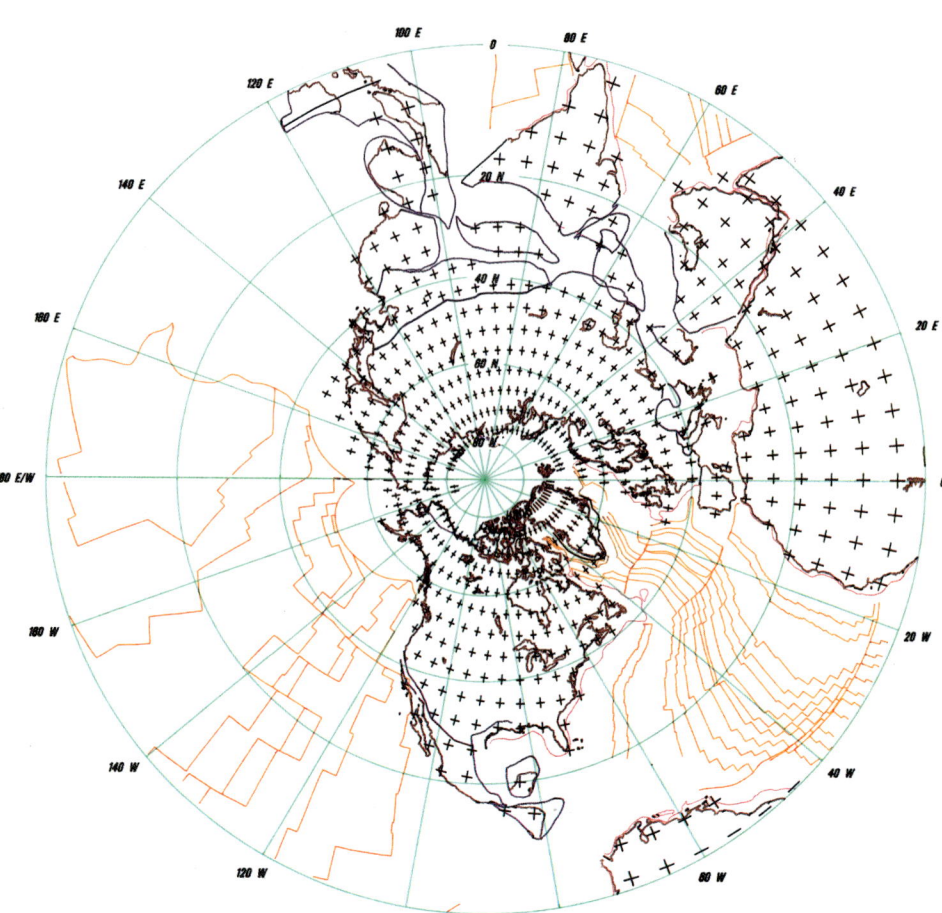

NORTH POLAR PROJECTION

POSITION OF LANDMASSES IN RELATION TO THE NORTH AND SOUTH POLES IN THE LATE TERTIARY, 20 million years ago

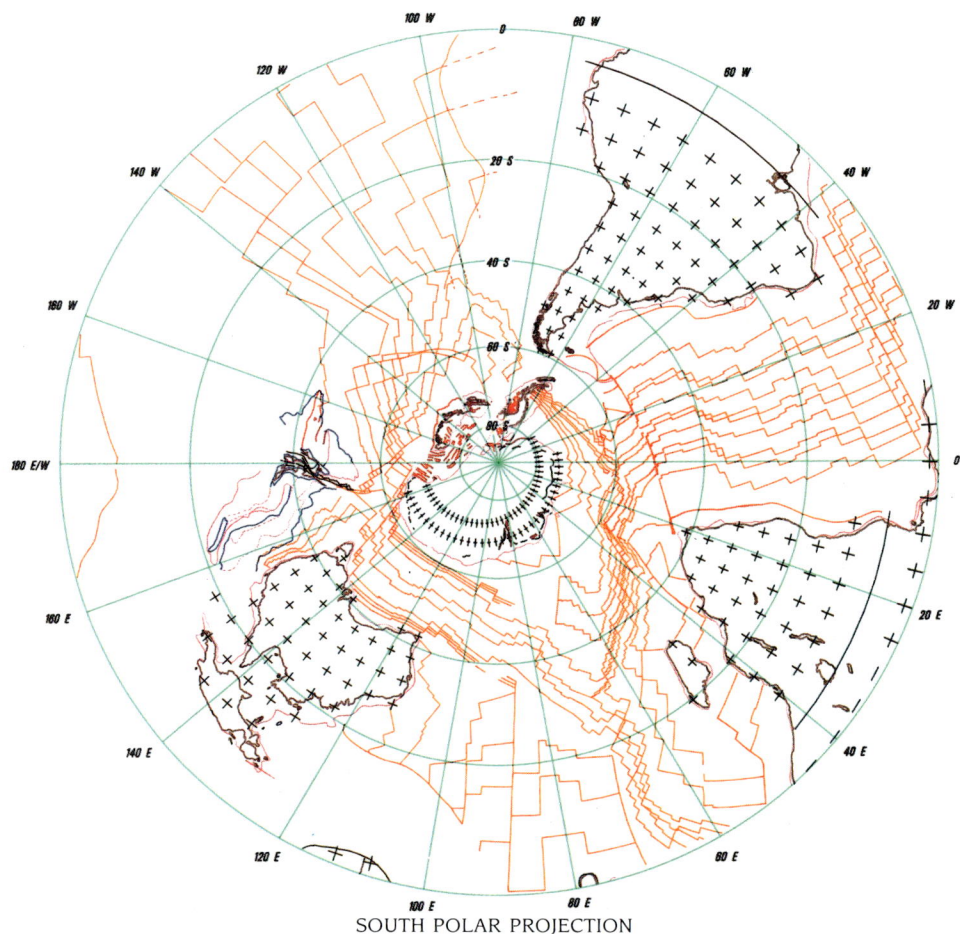

SOUTH POLAR PROJECTION

As the ice cap again increased later in the Epoch, sea levels dropped. Bass Strait opened in the Mid Miocene and was to close and open many times in the sea level fluctuations leading to, and during, the Pleistocene ice age.

With the formation of the permanent ice cap on Antarctica from 15 million years ago, temperatures declined and an increasingly dry anticyclonic circulation started the drying of central and northern Australia. River systems in arid Western Australia have been inactive since the Mid Miocene and are now reduced to a chain of salt lakes.

Throughout much of the Miocene there were Closed Forest communities in the southern part of the continent, much like those of the earlier part of the Tertiary. *Nothofagus brassii* was abundant. Laurels, Conifers and *Casuarina* were common around lakes and on river banks. Among the other vegetation were MYRTACEAE (including some with *Eucalyptus*-like pollen), PROTEACEAE and many other families. Forests like these extended to Alice Springs in the Centre.

Nothofagus brassii requires moderate rainfall and no dry seasons, and it declined sharply in the Late Miocene when climates changed and rainfall became seasonal as a result of the formation of the Antarctic ice cap and the continent's journey northwards.

Aridity increased, particularly in the north where the vegetation changed to forests dominated by MYRTACEAE. Grasslands appeared in some areas; Salt Bush (CHENOPODIACEAE) and *Acacia*, both indicators of dry conditions, appear in the Pollen Record at this time.

The Miocene has a good and ever-increasing Vertebrate Fossil Record in Australia. Footprints of giant Running Birds, the Mihirungs or Dromornithids, are known from Tasmanian rocks dated as Early Miocene. Some species of the genus *Genyornis*, which was abundant during the Miocene and Pliocene, are synonymous with the Aboriginal "mihirung

After the establishment of the permanent southern ice cap on Antarctica 15 million years ago in the Mid Miocene, aridity increased steadily in Australia.

A Skeleton of *Genyornis, a Giant Running Bird. This species was bigger than an Ostrich.*

DENSEY CLYNE

Ancestors of today's Emu lived in the Miocene Epoch. The spread of grasslands and Acacia scrub as the continent dried out suited them and other animals that had adapted to life on the open plains.

paringmal'' or ''giant Emus'' of the Dreaming as they overlapped with humans and may have become extinct not more than 6000 years ago. Broken shell from enormous eggs and bones of Birds as big as the Elephant Birds of Madagascar testify to the awesome size of some of these Birds. They were not ancestral to the Emu. Remains of fossil Emus also occur in the Miocene and it is notable that they were much smaller than those living today.

Fossils of Miocene Flamingoes are found in central Australia where they frequented the lakes. Another interesting Bird fossil is that of an Owlet-nightjar from the diatomaceous-earth deposit at Chalk Mountain at Bugaldie in New South Wales. It is related to Frogmouths, to the Potoos of South America, and to Nightjars. Also in the ''chalk'' where this fossil was found are impressions of Eucalypt leaves and gumnuts, probably of *Tristania*, and flowers of *Ceratopetalum*, the New South Wales Christmas Bush, along with other leaves.

The first land Mammal ever to have been described from Australia was the Miocene Diprotodontid *Wynyardia* known from only one skeleton found at Wynyard in Tasmania. It was, until quite recently, the oldest record of a Mammal in Australia, but is now much pre-dated by the *Steropodon* jaw from the Cretaceous (which is 110 million years old) and by the assortment of animals in the Etadunna Formation now known to be of Oligocene age.

Sites in central Australia have yielded Marsupial faunas, and the famous Riversleigh locality of Mid Miocene age (or Late Oligocene to Miocene — the date is not yet positively determined) in north-western Queensland is revealing a marvellous assortment of Bats as well as Lungfish, freshwater Tortoises, Crocodiles, huge Running Birds (Dromornithids), Diprotodontids, meat-eating Kangaroos and a diversity of smaller Marsupials, many new to science. The site is among the most important and significant fossil localities in the world.

Leaves of Eucalyptus *preserved in diatomite at Bugaldie, New South Wales. (Specimen AMF. Magn.X 1.0)*

A flower and a leaf of Ceratopetalum, *the New South Wales Christmas Bush, preserved in diatomite 15 million years old at Bugaldie. (Specimens AMF. Magn.X 3.0)*

DIATOMS

Diatoms are classified in the Division (Phylum) Thallophyta — the Algae — as they contain chlorophyll and photosynthesise carbohydrate in the manner of plants. Their golden pigments mask the green colour, hence their classification in the class Chrysophycophyta. They are microscopic unicellular (or acellular) organisms which secrete shells of silica comprising two parts, one fitting over the other like a lid.

Marine Diatoms appeared in plankton in the Cretaceous, while freshwater Diatoms appeared in lake deposits of Tertiary age. Their shapes and ornamentation vary endlessly, but marine forms tend to be radially symmetrical and freshwater forms are usually elongate.

Modern Diatoms are astronomically abundant in cool seas and their cases (called frustules) form a large part of the sediment which settles on the ocean floor. They are also abundant in lakes and streams and may even occur in damp soil. They require light for photosynthesis, so those in the seas are confined to the photic layers less than 180 metres deep.

Deposits of Diatomaceous Earth, resulting from the accumulation of untold billions of the microscopic shells, may be mined for use in filters, abrasives, fillers and other commercial purposes.

RON OLDFIELD

RON OLDFIELD

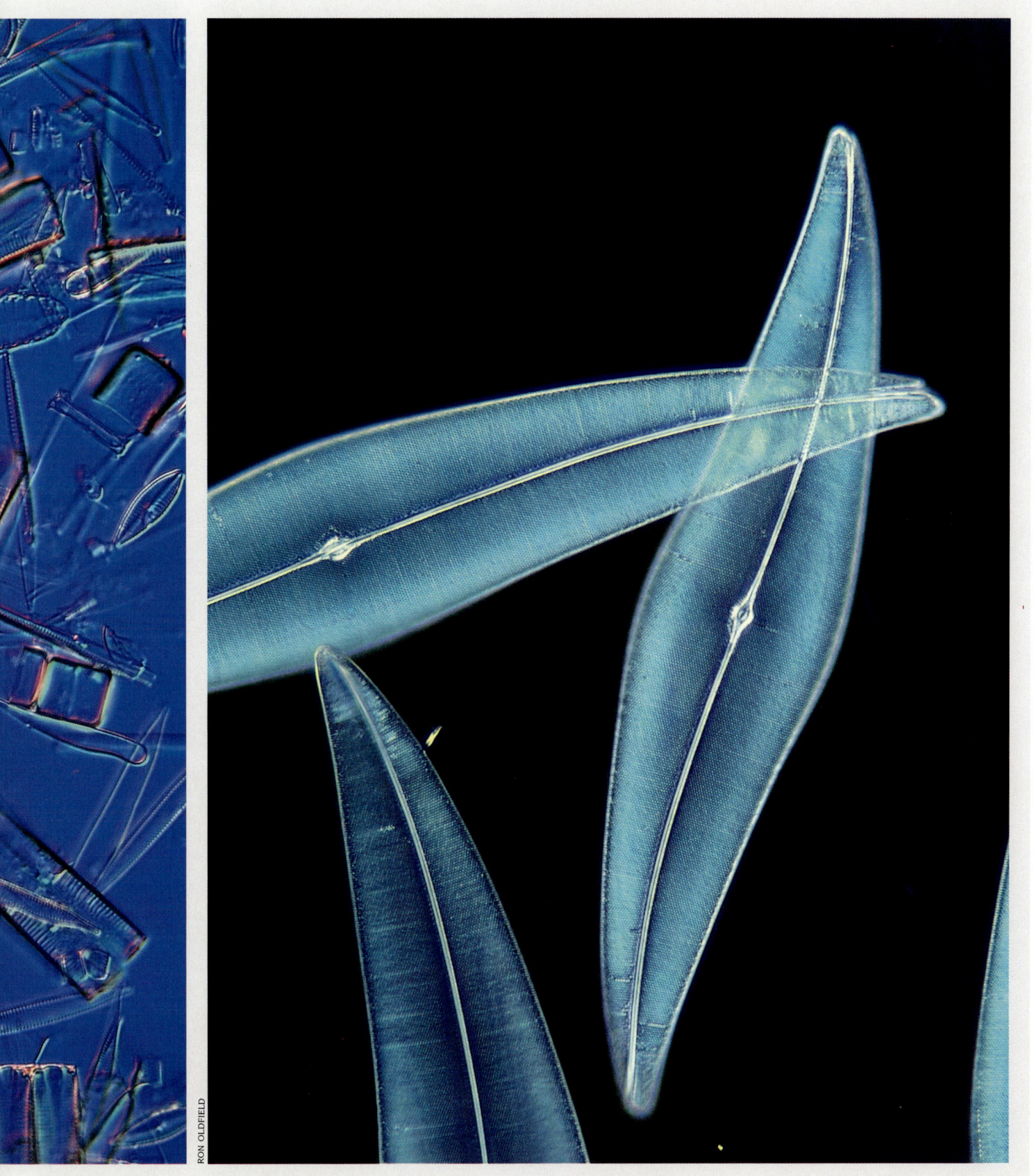

RON OLDFIELD

Thalassina, a Crayfish found in Tertiary sediments of Queensland and the Northern Territory. (Specimen AMF. Magn.X 1.6)

A freshwater Crayfish from Mt Wilson in the Blue Mountains of New South Wales.

Lovenia forbesii, a Tertiary Echinoderm from Victoria. (Specimens AMF. Magn.X 2.6)

The skull of a Diprotodontid, Neohelos, from the Miocene at Bullock Creek, Northern Territory. Neohelos was a Sheep-sized browser. (Specimen MV.)

DENSEY CLYNE

THE PLIOCENE

FROM 5.3 TO 1.6 MILLION YEARS AGO
DURATION: 3.7 MILLION YEARS

The "Terminal Miocene Event" with its plunging temperatures and the great expansion of the southern ice cap was followed in the Pliocene Epoch by a warming phase between 5 and 3 million years ago. Part of the West Antarctic ice sheet may have melted during this time. Renewed cooling followed, and the northern ice cap started to form 3 million years ago. The Arctic Sea was frozen for the first time during this Epoch.

The fluctuating patterns of colder/warmer, drier/wetter caused by the expansion and contraction of the ice caps became rapid pulses. The fullest extent of the northern ice was only about half the volume it would attain later during Pleistocene times.

The rising of the Isthmus Of Panama early in the Pliocene increased the cooling processes, as it formed another blockage to interrupt the equatorial currents. Australia's northwards movement resulted in the deformation of the regions to its north, elevating more land which made migration from Asia easy particularly when the continental shelves were exposed in glacial intervals. The arrival of Rodents in Australia dates from this time (about 5 million years ago) and the only Placental Mammals recorded before then were Bats, which had arrived in the Miocene (or Late Oligocene, if the Riversleigh deposits are proved to date from that time).

The trend towards aridity had begun in the Mid Miocene and the drying out of the Australian continent accelerated after the warm spell at the start

Gumnuts, fruits of Eucalyptus, from Pliocene mudstones at Gragin near Inverell, New South Wales. (Specimen MMF. Magn.X 3.5)

of the Pliocene, with the development of markedly different subregions and the spread of the central deserts. Ice age regimes became firmly established with the formation of the northern ice cap 3 million years ago.

Australia's south-western and southern margins received winter rainfall as anticyclonic high pressure belts moved north. The evolution of the highly individual southern Western Australian flora with its 80 per cent endemism of species started with the establishment of this Mediterranean-style climate. (In the south-western Cape in South Africa a similar climate developed at the same time, and parallel development of the flora there, from a similar Gondwanan stock, has resulted in another unique flora. Remarkable similarities exist between the floras of the two regions, which occupy the same positions on their own continents. The genera they contain are distinct because they have evolved in isolation, but at the family level and in the structure and content of the vegetation the common origins of the two floras and the like responses of their components to the same environmental pressures is strikingly demonstrated.)

The continuing plate movements had brought the northern part of Australia into a hot and humid climatic belt which counteracted the general trend towards desiccation of the whole continent. The elevation of the New Guinea mountains and the presence of high ground along the eastern margin of the Australian continent, where rainfall and conditions remained suitable, were important in providing refuges for Closed Forest and Rainforest vegetation. The refuge areas were also vital for the survival of the many Marsupials, Birds and other creatures which were adapted to life in such forests.

The Kosciusko Uplift, a gentle tectonic raising of the eastern highlands, caused rejuvenation of the river systems flowing to the coast and contributed to the formation of deep river valleys which would later become the "drowned valley" systems characterising modern coastal areas.

The nature of the forests had changed through the Tertiary but not until the Pliocene was there any large scale development of open forests with ground cover of grasses and ASTERACEAE (Daisy family). By the Pliocene, grasslands had taken over in the already arid areas. The appearance of grazing Marsupials — the "Antelope equivalent" Kangaroos, with the long hopping stride suited for life on the open plains — followed the development of this new habitat. MYRTACEAE were dominant in Pliocene forests but this does not necessarily indicate that modern-style *Eucalyptus*-dominated forests were generally established by then. The pollen types are more abundantly of *Baeckea*, *Backhousia* and *Tristania*. The *Eucalyptus*-like pollen probably comprised *Angophora*, *Syncarpia* and *Metrosideros* as well as *Eucalyptus*, and formed a less dominant element of the vegetation. *Acacia* trees and shrubs were becoming the dry country colonisers, and *Casuarina* (Desert Oaks) were adapted to arid areas.

Around the Pliocene–Pleistocene boundary, the aridity of the continent exceeded its present proportions, though actual desert conditions were not reached until about 200,000 years ago. Probably there had been some localised pockets of virtual desert throughout the Neogene, allowing a highly specialised desert flora and fauna to evolve. These desert-adapted forms were capable of spreading when large desert areas became available.

The Marsupial megafauna of the Pliocene Epoch has been much publicised, largely because it catches the imagination. The huge *Diprotodon* (which was to survive into Pleistocene times and overlap with the arrival of the Aboriginal people), Giant Kangaroos and Wombats, great Running Birds, an assortment of strange and wonderful creatures including Palorchestids with trunks, and many "normal" looking Marsupials, Birds and Reptiles populated this not-so-different Pliocene Australia.

PLIOCENE POLLEN GRAINS

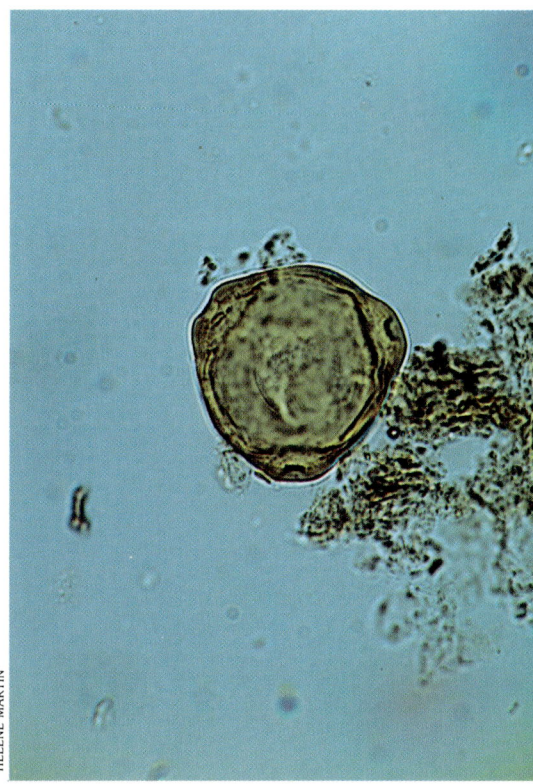

HELENE MARTIN

Pollen of CASUARINACEAE.

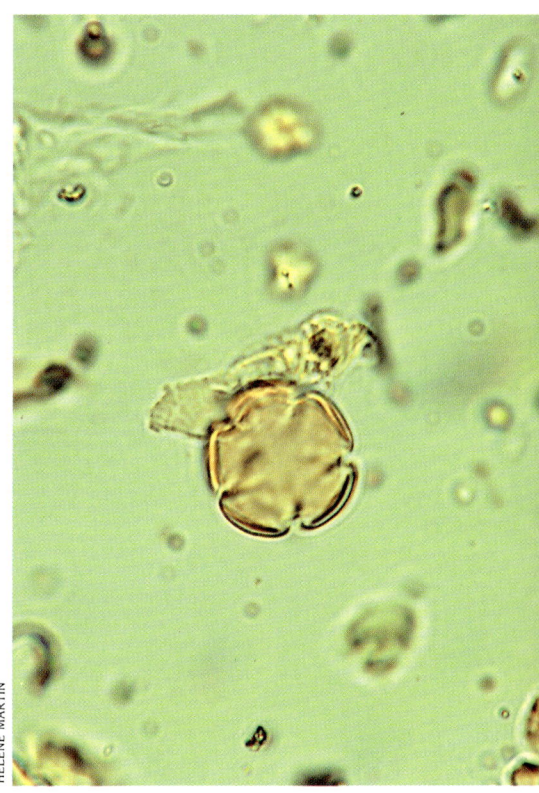

HELENE MARTIN

Pollen of Quintinia (SAXIFRAGACEAE).

From the Mid Miocene, when the onset of the ice age regimes coincided with the formation of the southern ice cap, the Australian fauna and flora had been forced to adapt or migrate in response to the general increase in aridity. Changing conditions were steadily converting the continent into the driest vegetated landmass on Earth. The Pliocene was a major time of transition between the widespread Closed Forest communities of the earlier Tertiary (which had begun contracting in the Mid Miocene) and the predominantly sclerophyll communities of the present day. The Pliocene bridges the gap between the ancient and the modern-style vegetation and climates, and only the fauna still presents a somewhat "prehistoric" aspect.

As the forests became open, the Closed Forest communities retreated to refuges including the fortuitously available New Guinea Highlands. There today, the closest approximation to the original ancient Rainforest Marsupial fauna tenuously survives, threatened by the destruction of the forests and the predations of protein-hungry New Guineans.

NEW ZEALAND IN THE TERTIARY PERIOD

By the start of the Tertiary, New Zealand's movement away from Australia was substantially complete. By 60 million years ago it had reached the position it now occupies in relation to the eastern Australian coastline — with approximately 2000 kilometres of Tasman Sea separating the two lands. New Zealand was a place of low relief, eroded down to plains supporting lowland forest of Gondwanan origin.

Throughout most of the Palaeogene the land was further eroded and the sea encroached on its margins, reducing it to a scatter of islands. But late in the Oligocene Epoch the trend towards submergence of the land was reversed, and the land started to re-emerge. During the Miocene and the Pliocene active tectonism saw mountains rising and the establishment of North Island and South Island in their modern form.

THE PALAEOGENE

FROM 66.4 TO 23.7 MILLION YEARS AGO
DURATION: 42.7 MILLION YEARS

The Palaeogene in New Zealand was a quiet time without tectonic activity and with only isolated volcanism until very near its end, in the Late Oligocene. The Ancestral New Zealand landmass had already been eroding throughout the Cretaceous Period and had become a lowland plain with little relief. Over the next 20 million years it was to undergo continued erosion as well as the inundation of its margins and lowest areas by the sea. Eventually the stage was reached where its survival as land was threatened by the erosion and subsidence, for it was reduced to a number of swampy low islands whose total area was about a third of that of modern New Zealand.

But at the end of the Oligocene Epoch the emergence of modern New Zealand began. On the boundary between the Australian and Pacific plates (along the orocline which had taken shape long before when Australia and

the outer edge of Gondwana had started their separation from Antarctica) the mountainous land started to rise. Activity on the Alpine Fault resulted in an enormous lateral movement, displacing land to the east of the fault from that to the west by an estimated 480 kilometres. The early movements were largely of a strike-slip type where the displacement of land on the two sides of the fault occurred simultaneously with the uplifting of the ranges, whereas the movement in the last 5 million years has been mostly elevation of the alpine mountains. It has been estimated that there has been 11 kilometres of uplift on the fault.

The flora and fauna which had entered Ancestral New Zealand as it emerged from the sea in the Late Jurassic, and the Cretaceous plants and animals which it had shared with the rest of Gondwana while it was still attached to the supercontinent, had been isolated for many millions of years before the Tertiary. The flora was a rich one of Flowering Plants, Conifers, Ginkgos, Cycads and Ferns, and the fauna included Running Bird ancestors, a Dinosaur-descendant ancestral to today's Tuatara, and Parrots.

The Kea, New Zealand's common large Parrot is descended from the Parrot Ancestor that was in Gondwana before the southern continents split up and New Zealand become isolated by the widening of the Tasman Sea.

THE PALAEOCENE EPOCH

FROM 66.4 TO 57.8 MILLION YEARS AGO
DURATION: 8.6 MILLION YEARS

During the Palaeocene the Ancestral New Zealand landmass was being eroded and little, if any, high ground remained. The extensive plains were forested and probably resembled the lowland forest areas of the Manawatu of North Island when European settlers arrived there 150 years ago. (The Maoris had by then already destroyed the coastal parts of this lowland forest, and the Europeans have since destroyed the rest with the exception of a few very small islands of forest which survive and serve to remind us of the glory that has departed and of the vandalism of Man.)

Sea levels were high and encroached on the margins of the land. Swamps were a feature of the lowland coastal plains, particularly in what are now the western approaches to the Cook Strait, and coals were formed from the vegetable matter produced in them. Drilling projects have established the presence of coal deposits offshore from Taranaki (Mt Egmont) and Nelson. The oil and natural gas of the Maui and Kapuni fields may have originated as a by-product of this Palaeocene coal formation.

The Palaeocene Epoch was warm with warm seas. The presence of Corals and *Globigerina* oozes (Foraminifera with calcareous shells) testifies to the warmth of seas. There is no Plant Macrofossil Record for the Epoch, though a rich and diverse flora is indicated by the Pollen Record. Podocarp Conifers and the southern Beech *Nothofagus fusca* were abundant. The family MYRTACEAE was represented by *Metrosideros* (modern Rata) and *Leptospermum* (Tea Trees). CASUARINACEAE, PROTEACEAE and *Anacolosa* (OLACACEAE) were widespread. The families TILIACEAE, LABIATAE and LAURACEAE are prominent and many others are represented. *Mauritia* and other Palms were abundant.

PALAEOGEOGRAPHY OF NEW ZEALAND
DURING THE PALAEOCENE EPOCH

THE EOCENE EPOCH

FROM 57.8 TO 36.6 MILLION YEARS AGO
DURATION: 21.2 MILLION YEARS

Continued erosion of the land and further encroachments of the sea were to reduce the area of Ancestral New Zealand by half during the Eocene Epoch. By late in the Eocene most of the land was barely above sea level — a low, swampy place eroded down to flat plains with no significant relief, but still enjoying a temperate climate and no shortage of rain. It seems likely that weak and variable atmospheric circulation patterns prevailed.

The coal swamps at the western approaches to the Cook Strait region, which had been a feature of the Palaeocene, persisted. Late in the Epoch the sea invaded the region, separating the Taranaki and Nelson coal swamps by an intervening stretch of ocean. The main coalfields of the Westport–Greymouth districts were produced from organic matter which collected in the swamps of the west coast of South Island. (Eocene coals have been detected also on the Campbell Plateau between New Zealand and Antarctica. Some low-lying ephemeral land must have existed there and it may be significant in the context of assessing New Zealand's isolation. The nature of the plateau and of the rises in the area between Australia and New Zealand may imply that some island chain connection was possible for migrating plants early in the Tertiary, though this suggestion is highly speculative.)

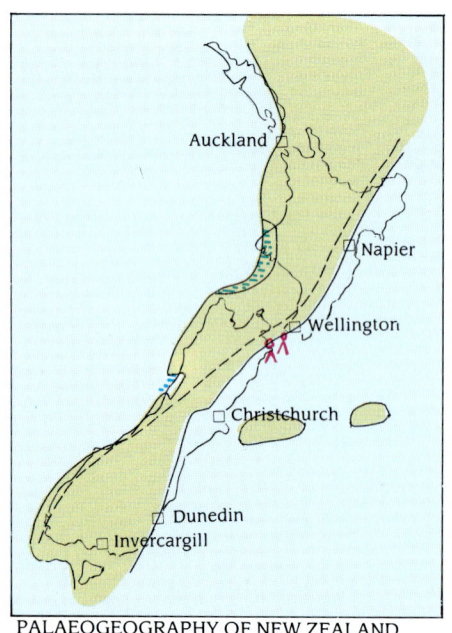

PALAEOGEOGRAPHY OF NEW ZEALAND
DURING THE EARLY EOCENE EPOCH

About 40 million years ago the sea invaded the Fiordland area of South Island, from the Waiau Valley to the Hollyford, and reached far inland in Marlborough. Marine incursions into Canterbury almost met the sea encroaching from the west coast, leaving only a narrow isthmus of land. Large swamp areas near the top of North Island produced the Waikato coal deposits which are being mined today.

Volcanic activity started at East Cape and Oamaru in South Island in the Late Eocene and continued into the Oligocene Epoch.

The composition of the vegetation as seen in the Pollen Record had changed slightly since the start of the Tertiary, with the proportions of some of the major components altering. The Beeches had become dominant, supplanting the Conifers. At first *Nothofagus fusca* was the most abundant species, then *Nothofagus brassii* group pollens took over and remained the major component. *Phyllocladus*, the Celery-top Pine, and *Cupania* (SAPINDACEAE), EUPHORBIACEAE and MALVACEAE (*Hoheria*) were present as in the Australian Eocene flora. *Elytranthe* (a Mistletoe) became abundant as its host trees, the Beeches, gained dominance. The PROTEACEAE are represented by pollen of *Banksia* and *Hakea* type, and a number of other genera, possibly including *Knightia* which is one of only two genera of the family which survive in New Zealand's modern flora.

In the Late Eocene the Flax *Phormium* (AGAVACEAE) appears in the Pollen Record with *Freycinetia*, the only genus of the PANDANACEAE which today occurs in New Zealand. *Astelia* (LILIACEAE), *Quintinia* (SAXIFRAGACEAE) and the *Rhopalostylis* Palms make their first appearance. Other taxa in the Eocene Pollen Record which are shared with the Australian flora of the same age are: *Ilex* (AQUIFOLIACEAE), SAPOTACEAE, BOMBACACEAE (Baobab family), CRUCIFERAE, ARALIACEAE, EPACRIDACEAE, MELIACEAE and PAPILIONACEAE.

There was little diversity in the vegetation types over New Zealand during the Eocene, as would be expected when the landscapes were all plains and swamps and the climate uniformly warm-temperate. The cooling episode at the Eocene–Oligocene boundary may have caused the extinction of some of the "tropical" and "subtropical" elements, including a number of the Proteaceous types.

Eocene marine Invertebrate faunas were rich in Brachiopods and Molluscs, and Corals indicate that the seas were warm throughout most of the Epoch.

Embothrium coccineum, a South American member of the family PROTEACEAE, *whose closest relative is found in Australia.*

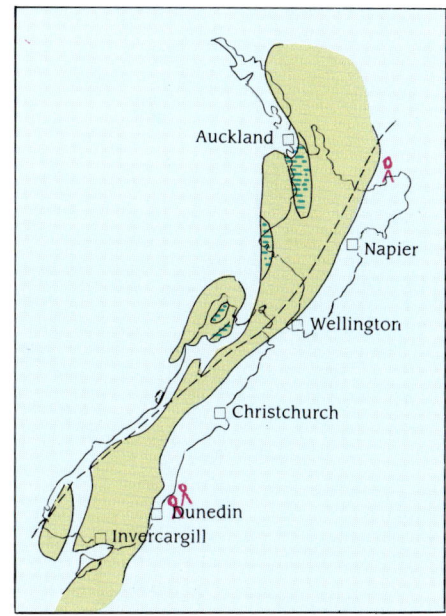

PALAEOGEOGRAPHY OF NEW ZEALAND DURING THE LATE EOCENE EPOCH

The cooling in the Late Eocene is confirmed by the presence of fossils of Giant Penguins, at the same time as they appear in the record in Australia and in Seymour Island in the Antarctic.

THE OLIGOCENE EPOCH

FROM 36.6 TO 23.7 MILLION YEARS AGO
DURATION: 12.9 MILLION YEARS

The eroding away of the land and its gradual submergence which had started in the Late Cretaceous continued during the Oligocene, until the area of land which remained above the sea (when the areas of all the small island fragments were added together) was only less than a third of the area of modern New Zealand. These fragments of land formed a "changing archipelago" of small islands with the two largest at the northern and southern ends.

The phase of erosion and gradual submergence reached its maximum in the Mid Oligocene. At the climax of the submergence, the Eocene coal swamps of North Island were inundated. Deep seas covered the entire eastern flank of the land and cut it in two in the National Park region, crossing the Alpine Fault zone where the mountains are today.

South Island was even more thoroughly submerged. Only the old eroded plains of Canterbury and Otago survived reasonably intact. Coal swamps formed in low-lying areas in Southland (the Gore Coalfield) and in Otago. The Fiordland region was almost completely submerged and the Hollyford–Waiau region became a deep marine trough. A chain of volcanic islands erupted in a line through Marlborough and into inland Canterbury. The volcanoes of East Cape and Oamaru, which had become active in the Eocene, continued to erupt.

In this ravaged land the forests still grew and the animals continued to live in them. It is amazing how much of the flora and fauna managed to survive even up to this point in New Zealand's history. In the forests the Beech *Nothofagus brassii* dominated. CASUARINACEAE was often abundant, probably because it was well suited to the sandy water margins and disturbed areas of the times. MYRTACEAE, Palms, Podocarp Conifers and *Nothofagus fusca* were other prominent components of the vegetation.

A fossil flora from Southland and another from the west coast of North Island have rather different assemblages of plants. The Southland flora is rich in *Nothofagus brassii* and Ferns, with *Nothofagus menziesii* and a Podocarp Conifer *Microcachrys* being common. Also present are *Weinmannia* (CUNONIACEAE) which is the "Kamahi" of modern forests, *Pseudowintera* (WINTERACEAE), RANUNCULACEAE, UMBELLIFERAE and others. In some pollen samples an abundance of *Mallotus* (EUPHORBIACEAE) occurs, associated with Lilioid and Palm pollen. This assemblage suggests a drier coastal environment similar to that on parts of the Queensland coast of Australia today. In contrast to the diverse flora of Southland, the North Island flora was 80 per cent dominated by the Beech *Nothofagus brassii* and there are few associated species.

The Beech-dominated floras of the Oligocene had an admixture of other plants whose present-day distribution is interesting. Among these are the proteaceous genera *Isopogon* and *Embothrium* now absent from New Zealand but present in Australia, the EPHEDRACEAE now absent from both New Zealand and Australia, and the BOMBACACEAE now absent from New Zealand while some (Baobabs) still exist in a very limited area in northern Western

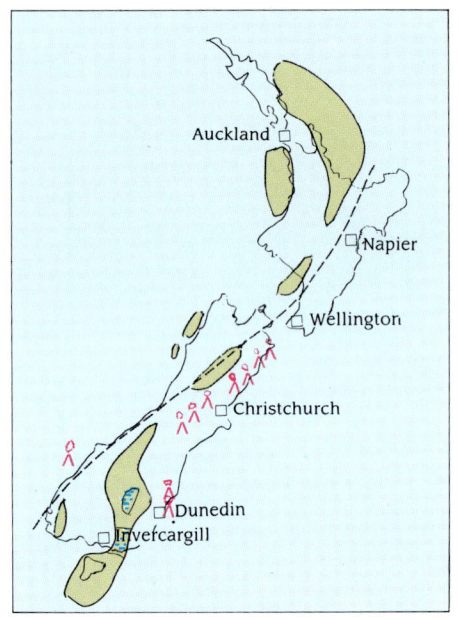

PALAEOGEOGRAPHY OF NEW ZEALAND IN THE OLIGOCENE EPOCH

Australia. (*Ephedra* is a plant which occupies a position halfway between Flowering Plants and Gymnosperms, and although now absent from both New Zealand and Australia, it still survives in many parts of the world.) RESTIONACEAE, SPARGANIACEAE (Burr Rushes) and TYPHACEAE were marsh plants of the Oligocene and persist to this day. The "Mapou" or *Myrsine* (MYRSINACEAE), *Astelia* (LILIACEAE), *Coprosma* (RUBIACEAE), *Laurelia* (LAURACEAE) and *Fuchsia* (ONAGRACEAE) are all still part of modern New Zealand's flora.

Invertebrate faunas of the New Zealand Oligocene are rich in Alcyonarian Corals, and there are also famous Brachiopod faunas and abundant Echinoderms and Molluscs. Vertebrate fossils comprise Penguins and Whales (Zeuglodonts and Squalodonts).

Near the end of the Oligocene Epoch the submerging trend which had so reduced the land was reversed and a phase of emergence began. The Kaikoura Orogeny re-activated the Alpine Fault, and lateral movement started to displace the land on its eastern and western sides. Even by latest Oligocene times some topographical relief was evident, particularly along the centre of North Island.

THE NEOGENE

FROM 23.7 TO 1.6 MILLION YEARS AGO
DURATION: 22.1 MILLION YEARS

New Zealand became the North Island and the South Island which we know today during the Neogene. The Miocene and Pliocene Epochs saw great activity on the Alpine Fault, thrusting up the alpine mountains which characterise modern New Zealand and displacing land on one side of the fault from that on the other.

THE MIOCENE EPOCH

FROM 23.7 TO 5.3 MILLION YEARS AGO
DURATION: 18.4 MILLION YEARS

From the beginning of the Miocene, rapid emergence of the land proceeded throughout New Zealand. In the centre of North Island rising axes led to marked changes in the nature of deposition of sediments. As the elevated land was eroded many small basins became filled with sediment. The rising ridge through Northland and Auckland was flanked by a chain of andesite volcanoes and the first eruptions in the Coromandel Peninsula began. On the west coast, deltas in northern Taranaki supported coal swamps which were to produce the Mokau coalfield.

By the Late Miocene New Zealand's land area was about double what it had been at the start of the Epoch. For the first time since the Late Cretaceous, when the Ancestral New Zealand landmass had boasted a mountain range, mountains and hills appeared — in Marlborough, south Nelson, south Wellington and south Hawkes Bay. The Alpine Fault and associated smaller faults were highly active, displacing the east from the west laterally, and fault-bounded blocks began to rise independently of others, creating the beginnings of the modern alpine mountain chain which would grow as the Kaikoura Orogeny became fully operational.

A chain of volcanic islands was active off the north Taranaki coast. In South Island, volcanic activity started at Lyttleton and in the Otago Peninsula, and was to continue until the Pliocene.

The Early Miocene vegetation differs from that of the Oligocene in the

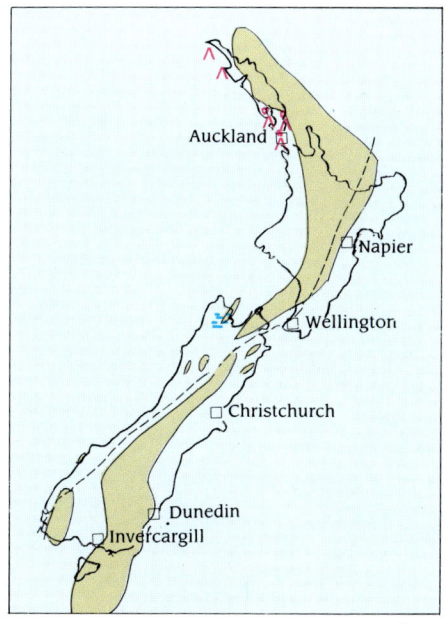

PALAEOGEOGRAPHY OF NEW ZEALAND
DURING THE EARLY MIOCENE EPOCH

The Kiwi, symbol of New Zealand, is a descendant of the Giant Running Bird ancestor that entered the land when it was still connected to Gondwana. Another line of descent from the same ancestor lead to the Moas which dominated the fauna and filled most niches in the absence of Mammals. The Moas only became extinct when hunted systematically off the face of the land by the Maoris.

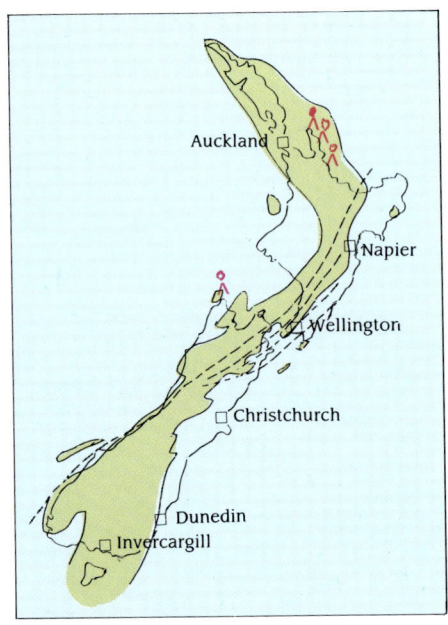

PALAEOGEOGRAPHY OF NEW ZEALAND
DURING THE LATE MIOCENE EPOCH

relatively greater abundance of Podocarp Conifers, Ferns and Palms, and in an increase in the percentage of *Nothofagus fusca*. The dominant forest trees were still Beeches of the *Nothofagus brassii* group. In Southland conditions were once again drier, and there was a decrease in Podocarps and Ferns. At Coopers Beach in Northland there is a well-known fossiliferous bed containing fruits of a Coconut Palm, *Cocos zeylandica*. The nuts are very small compared with living species.

Miocene floras in New Zealand include: the Monocotyledons *Cordyline* (Cabbage-trees), *Bulbinella* (LILIACEAE), and *Potamogeton* (POTAMOGETONACEAE) pondweeds; and the Dicotyledons *Alectryon* (SAPINDACEAE), CHENOPODIACEAE (Salt Bushes), CORNACEAE, *Dactylanthus* of the family BALANOPHORACEAE (in which only one species survives today, confined to New Zealand), *Halagoris* (of the family HALAGORACEAE, and a genus which is abundant in modern Australia), "Kawakawa" or *Macropiper* (of the family PIPERACEAE, and which grows as an understorey in modern forests), *Ixerba* (which is an endemic genus of New Zealand today and belongs to the family SAXIFRAGACEAE), *Pittosporum* (PITTOSPORACEAE), *Muehlenbergia* (POLYGONACEAE), UMBELLIFERAE (also called APIACEAE), and "Mahoe" or *Meliclytus* (VIOLACEAE).

The family AQUIFOLIACEAE, represented by *Ilex* Hollies, becomes extinct in the Miocene in New Zealand, as do *Dacrydium* with pollen of the *franklinii* group, and MYRICACEAE and a few others.

The latest part of the Miocene shows a vegetation change in response to the cooling of the Terminal Miocene Event. Shrubby ASTERACEAE of the Tubuliflorae section of the family become locally abundant with the appearance of heath zones in response to the cold on the higher ground.

The warm climate of the Early to Mid Miocene with its warm seas is reflected in the nature of the abundant Foraminfera and the Corals in marine formations. The sudden cooling when the southern ice cap formed in the Mid Miocene, 15 million years ago, eliminated some warm-water forms from the Fossil Record. The Brachiopods, which had undergone explosive development in the New Zealand Oligocene, also decreased in importance as the waters cooled.

The Late Miocene is marked by the first record of Moas and the footprints of a Kiwi.

THE PLIOCENE EPOCH

FROM 5.3 TO 1.6 MILLION YEARS AGO
DURATION: 3.7 MILLION YEARS

Tectonic movements of the Kaikoura Orogeny became even more intense in the Early Pliocene, and modern New Zealand was fast taking shape. The faults on the Alpine Fault System were all active, pushing up blocks which were creating the main ranges. South Island was almost its present shape, with marine deposition only on the eastern margins of Canterbury and Marlborough, and in small areas of coastal Westland. The Taranaki–Wanganui, Gisborne–Hawkes Bay and southern parts of the Wairarapa areas of North Island were submerged early in the Pliocene. Some melting of the huge southern ice cap, which had formed in the Terminal Miocene Event, resulted from global climatic warming and caused a sea level rise. The flooded areas lay to the north of the present Cook Strait, which was bridged by a peninsula of land extending from a northern extension of the Nelson–Marlborough margin.

Marine straits crossed the rising ranges, forming the Manawatu Gorge and the Kuripapango in inland Hawkes Bay. Subsequent tectonic movements resulted in the seas withdrawing and all the land reappearing. Widespread volcanic activity occurred in Northland, the Coromandel Peninsula and Waikato throughout the Pliocene. In the extreme north, the stack-like islands of the North Cape eruptives became welded onto the main Northland Peninsula and a giant tombolo (sand spit) was deposited on the junctions, connecting the islands to the mainland above sea level.

Twin basalt volcanoes erupted and produced high cones on what is now the Banks Peninsula on the east coast of South Island. Today the eroded craters of the now-dead volcanoes are the sites of Lyttleton and Akaroa harbours. Port Chalmers on the Otago Peninsula is another harbour formed from the crater of a basalt volcano. There is a strange kind of irony in these safe havens existing in places which have seen such elemental violence in the past.

The tempo of climatic fluctuations which had begun with the formation of the southern ice cap in the Mid Miocene, 15 million years ago, was accelerated by the formation of the northern ice cap some 3 million years ago in the Pliocene. Fluctuations in the vegetation types corresponding to the warmer/wetter, colder/drier cycles of the ice age resulted.

In the warm interval at the start of the Pliocene Epoch *Nothofagus brassii* dominated the forests in North Island and other plants formed only a minor component of the vegetation. In South Island the forests were more mixed, with MYRTACEAE (Rata) prominent, along with *Dacrydium cupressinum* (Rimu or Red Pine) and *Nothofagus fusca*. Palms were an important constituent of the flora in the warm interval, and have not been a significant element since. Tree-ferns were more abundant and diverse than they are now.

In a cold/more arid fluctuation at about 3 million years ago when the northern ice cap formed and sea levels were low, the vegetation in coastal lowlands in the north of South Island was grassland with abundant ASTERACEAE, UMBELLIFERAE, *Euphrasia* (SCROPHULARIACEAE), and CYPERACEAE (Sedges). Environments were obviously changing rapidly with the climatic fluctuations, and there were many different sorts of plant communities — a situation that would be expected in a land with high mountains, uplands and plains experiencing the fickle changes of an ice age. The overall forest type was Beech/Podocarp, with a mix of cool-temperate taxa.

The Late Pliocene was the last time when *Nothofagus brassii* Beech was to be prominent. Its importance declined during the ice age, presumably because the fluctuations from arid to wet were not to its liking. MYRTACEAE took over the more dominant role, replacing *Nothofagus brassii* in South Island communities.

Under the ice age conditions many of the warm-temperate taxa disappeared. They had nowhere to flee to in the small, isolated island. This situation contrasts with that in Australia where plants had room to move and find their preferred conditions as climatic change affected them. Thus in New Zealand *Bombax* (Baobab family), various members of the PROTEACEAE, *Ephedra*, *Zygogynum* (WINTERACEAE), two species of *Cupania* and several other taxa did not survive beyond the Pliocene. *Acacia* became abundant in arid coastal sites and a large number of herbs and plants of open heaths appear in the Pollen Record. These include *Pimelia* (THYMELEACEAE), PORTULACACEAE, *Epilobium* (ONAGRACEAE), *Colobanthus* (CARYOPHYLLACEAE), *Wahlenbergia* (CAMPANULACEAE), *Arthropodium* (LILIACEAE), *Gentiana* (GENTIANACEAE) and *Hebe* (SCROPHULARIACEAE).

The marine Invertebrate record shows that several warm-water genera became extinct in New Zealand waters as the colder conditions intensified. Late in the Pliocene the pelagic Mollusc *Hartungia postulata* appeared. It has close relatives in the Pliocene of the Azores, Morocco and Australia.

A Pliocene Seal, *Arctocephalus*, is an important Vertebrate fossil which was found near Cape Kidnappers.

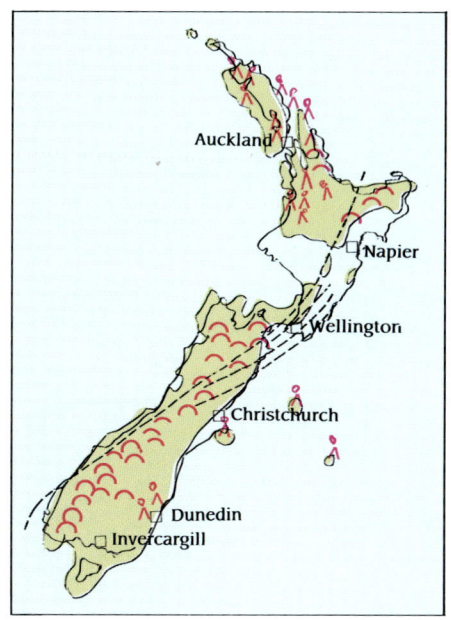

PALAEOGEOGRAPHY OF NEW ZEALAND IN THE PLIOCENE EPOCH

OVERLEAF:

HUNTING THE DIPROTODON
About 30,000 years ago

The Short-faced Giant Kangaroo in the background and the large, amiable *Diprotodon* were no match for Aborigines hunting with spears and fire. The vegetation of the riverine environment comprised Paperbarks, Casuarinas and Grass-trees just like today, but many of the Marsupials were "megafauna" — far larger than the species alive now. The extinction of the giants is believed to be the result of human hunting activities, and the later extinction of many smaller species relates to the introduction of the Dog (Dingo) by Aborigines.

HUNTING THE DIPROTODON

CHAPTER 10

WHEN THE ICE CAME AND WENT

THE QUATERNARY PERIOD
THE LAST 1.6 MILLION YEARS

The Quaternary Period is divided into the Pleistocene Epoch, which comprises the bulk of the Period and was characterised by the repeated glaciations of an ice age, and the Holocene Epoch, which represents the last 10,000 years since the retreat of the most recent ice sheets in the Northern Hemisphere. The divisions of the Period are somewhat arbitrary, as the ice age began well within Pliocene times, and glaciation has not disappeared in the Holocene. The North Pole and South Pole today remain within an ice age.

The main direct effects of the ice age were felt by continents in the Northern Hemisphere where land is closely gathered round the North Pole. For much of the rest of the world, aridity was more significant than intense cold, and this was the case in Australia. The fluctuations in sea level associated with the increase and decrease of the ice caps affected the shores of all lands.

The Quaternary histories of Australia and New Zealand differ greatly. Australia was a tectonically stable landmass little affected by glaciation, while New Zealand suffered partial submersion, convulsive and dramatic mountain-building movements with volcanism and earthquakes, and also widespread and severe glaciation.

Trace fossils of early Aboriginal occupation of Australia: hand prints, about 20,000 years old, on a cave wall in the Northern Territory.

Pleistocene Coral from Evans Head, New South Wales. This Acropora Coral extended its range further south in previous interglacials than it does today. (Specimen MMF. Magn.X 1.6)

The "Great Ice Age" is now known to have been well under way by 3 million years ago, so it is no longer considered as only a "Pleistocene Ice Age". The fluctuation from severe cold to brief warmer times has been repeated possibly 30 times since the Mid Pliocene. Ice age regimes, of course, go back far in time. They include a cold episode about 35 million years ago at the end of the Eocene Epoch, the formation of the southern ice cap at 15 million years ago in the Miocene Epoch, and the cold event which was part of the Terminal Miocene Event at 6 million years ago. The formation of the North Polar ice cap at 3 million years ago in the Pliocene was the final act, establishing the "Great Ice Age" which persisted and characterised most of the Quaternary Period. The deglaciation since the last glacial peak at between 16,000 to 14,000 years ago has been the feature of the Holocene and the lead up to it.

Because of the effect of this ice age on the evolution of the human species, and on the lifestyle of people even within the historical past, our perspective on it is more personal. The rapid rise of Modern Man to such super-dominance — and, indeed, plague proportions — would not have been able to take place so easily had the last 10,000 years not been a benign interglacial. Environment has been the spur to technological and cultural evolution of the human species, just as it has been significant in the adaptation of other species over far longer geological time spans through the ages.

The "normal" state of the world through the billions of years of geological time seems to have been one of ice-free Poles and smaller seasonal contrasts than we experience today. Ice ages have interrupted the warm conditions about every 250 million years, and each ice age in the Phanerozoic Eon has lasted for about 50 million years. This history might suggest that the ice age in which our present interglacial falls is still with us, and there will be a return to glacial times followed by interglacials for perhaps another 15 million years. When the next deep-freeze will happen is anyone's guess. Expert estimates range from a frozen year 3000 AD to perhaps as far away as 23,000 years in the future, the latter taking into account the increasing man-made Greenhouse Effect which logically would postpone the next glacial phase.

A brief account of some of the Quaternary events in other parts of the world during the ice age provides a background to the situation in Australia and New Zealand:

The continents of the Northern Hemisphere bore the brunt of the ice, being grouped around the North Pole. (In the previous major ice age which was in the Late Carboniferous to earliest Permian times, it was the southern supercontinent which was most affected, as the South Pole was then on Gondwana and the North Pole was isolated by sea. In that ice age more than half of Australia's land surface was subjected to glaciation.) The northern ice cap reached its greatest extent about 16,000 years ago and was then up to 3 kilometres thick in parts. It is estimated that there were 40 million cubic kilometres of ice on the land. The ice covered Canada and extended down to New York and the Ohio River, and down the Rockies and Cascades. It spread over Scandinavia to Britain and froze over the region where London is now, and it covered Germany and European Russia. Round the iced-up regions were treeless plains with melt-water rivers south of the ice margin where reindeer roamed and hardy human hunters followed. South again of the bleak plains were the Boreal forests of Birch, Pine, and Spruce. Broad-leaf forest had retreated southwards from Europe to North Africa and the Near East, and in America to Florida and the southern edge of the Gulf States. Dry corridors caused gaps in the ice cap in Siberia and North Alaska.

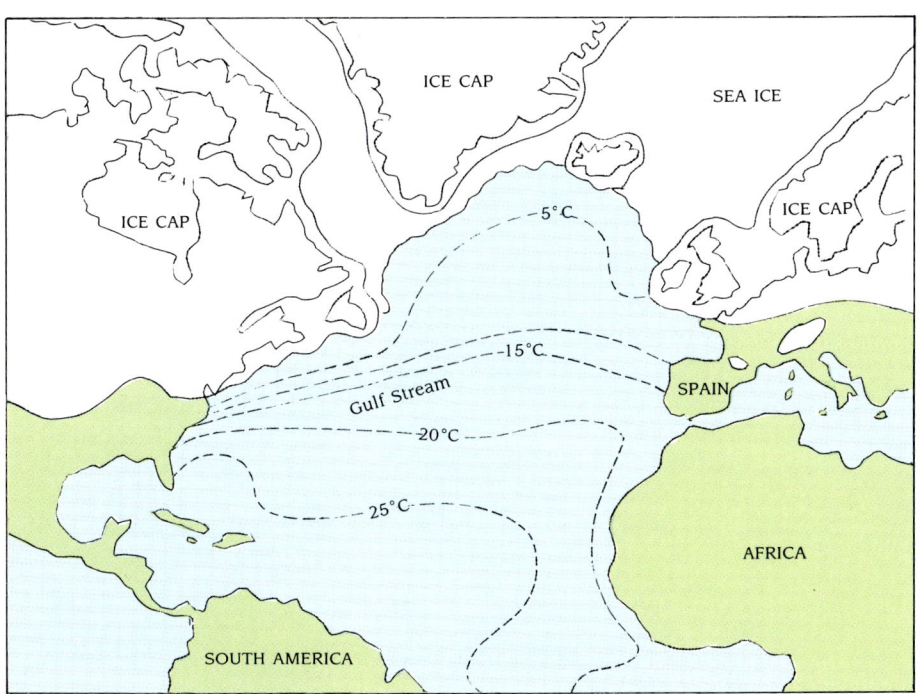

EXTENT OF ICE IN THE NORTHERN HEMISPHERE
At the last Glacial Maximum 18,000 years ago

The climate near the edges of the ice was not always as severe as might have been expected. Forests grew near the edges of some of the ice sheets, particularly in America where summer days were bright and sunny and winters were not truly arctic. The ice had been able to spread so far because of its own weight and spreading ability. It could spread faster than it could melt. The enormous volume of the northern ice cap was the result of its moving across land. (The South Polar ice cap is only centred on the Antarctic landmass and its edges enter deep water, so that the floating ice shelf is trimmed by waves and warmer water.)

During the ice age the North Atlantic was an Arctic sea. It was frozen over between Greenland and Britain, and there was sea-ice between Cape Hatteras and Spain. The Gulf Stream — a current of warmer water — crossed straight from Florida to Africa. The North Pacific was not as profoundly affected, though the cooler equatorial currents meant that many seas which are tropical today had water temperatures approximating those of the present-day North Sea in summer.

As the ice melted in Europe and Britain after the deep-freeze, Birch and Pine woods again replaced the tundra, and by 11,000 years ago the climate was quite mild. Then a sudden fluctuation over an interval of only 100 years, at about 10,500 years ago, brought the tundra back to Britain and the Low Countries, and the Reindeer and their hunters again held sway. This little ice age lasted for about 700 years, and then conditions again improved. About 9000 years ago European summers were almost as warm as they are now. Forests of Oak and Hazel covered England and Germany where only tundra had existed 1000 years before. This optimal climate lasted for about 4000 years, after which another little ice age occurred heralding more variable and generally cooler and wetter regimes. Greenland has never left this great ice age, neither has Antarctica.

Within recorded history there was a "little ice age" between the years 1650 and 1850 AD which made parts of Europe very unpleasant places to live.

By comparison with the Northern Hemisphere, southern lands were much less affected by the ice age. Australia was particularly fortunately located and missed most of the problems directly associated with ice and snow. The changes that occurred were caused by fluctuating climatic patterns, increased aridity and exposure of the continental shelves in cold phases. New Zealand was not so lucky and South Island experienced severe glaciation.

The skull and jawbone of a Wombat, Vombatus sp. cf. ursinus, very similar (and possibly identical) to the Common Wombat. Skull length 185 mm. Wombats are more closely related to Koalas than to other Marsupial groups. All Wombats are burrowers, so their pouches have rear openings to prevent them from getting full of sand when they dig. They have Rodent-like teeth with continuous growth, because their preferred diet of roots means that they chew sand and grit which rapidly wears down their teeth. From Naracoorte, South Australia. About 35,000 years old.

The arid Centre, with its "Spinifex" and other plants as well as its animals adapted to living under the harsh conditions that apply there, is the result of the extreme aridity of some of the glacial phases of the Pleistocene ice age.

AUSTRALIA IN THE QUATERNARY PERIOD

The Australian Quaternary system contains a variety of deposits. There are sediments which accumulated on the continental shelf, those that were spread by rivers, others which were produced by glaciers grinding their way across the landscape, and vast desert dunes blown into ridges by the wind. The diversity of deposit types is in accordance with all the different aspects of ice age environments. Continental shelves were alternately exposed and submerged, glaciation occurred on high ground, and arid phases alternated with pluvial periods when there was erosion. The sedimentary rocks tell the story of the changing times.

Australia was tectonically quiet by the Quaternary, and there were only two regions in which volcanic activity occurred. These regions were in northern Queensland and in part of Victoria and adjacent South Australia. Basalt flows of Pliocene to Holocene age cover about 13,600 square kilometres in the former, and more than 26,000 square kilometres in the latter. The most recent volcanic eruptions occurred at Mount Gambier in South Australia about 4850 years ago and at Tower Hill in Victoria about 4500 million years ago. Dingo bones and an Aboriginal grinding stone have been found below volcanic rocks at Tower Hill, so Man and his Dog were witness to the active volcanoes in this region.

Quaternary marine deposition occurred on the continental shelf off the Eucla and Canning basins, and some marine sediments were deposited in southern basins and in the northern Carpentaria Basin at times of high sea level. Predominantly carbonate sediments accumulated on the Sahul Shelf seawards of the Bonaparte Basin in Western Australia. Carbonate deposition also had been occurring since Miocene times and now continued on the major platform area off the Queensland coast where the modern Great Barrier Reef started to grow from about 8000 years ago when sea levels had stabilised. The adjacent Queensland Plateau had been the largest area of the world up until that time to support reef. It had died when sea levels on the plateau rose faster than could be accommodated by reef growth. The dead Coral on the Queensland Plateau is a stark reminder of the dangers of sea level changes, and should give us reason to worry about the future of the Great Barrier Reef. If the Greenhouse Effect is allowed to develop unchecked, there will be a sea level rise of great rapidity when seen in the context of geological time, and it is quite likely the Coral will be unable to grow fast enough to keep up with it. Then one of the great natural wonders of the world will die.

Throughout the Quaternary, surface-type deposits forming sand and soil cover of the land were spread by rivers. In particular, the inward drainage system of the Lake Eyre Basin accumulated such deposits from the rivers which drained into it. The Channel Country today with its mesh of rivulets, its sand and its mud, and the alternating cycles from a usually desiccated state to a floodplain, continue the Pleistocene regimes of spreading sediment. Rivers feeding into other basins also spread sediment. Aridity, and the consequent decrease in vegetation, and alternating wetter times led to rapid erosion. As aridity increased, blown dunes became a feature of landscapes. The parallel dunes of the central deserts date from a period of intense aridity about 300,000 years ago, and mobile dunes have been characteristic of the arid places ever since. The sand-ridge deserts of the Canning and Lake Eyre basins, in particular, are amazing landscapes. The ridges are dunes from 12 to 30 metres high, separated by corridors between 250 and 450 metres wide. The dunes themselves are narrow, rectilinear and tens of kilometres in length — some have been measured and found to

ALLAN WELLS

exceed 300 kilometres. They often converge, producing assymmetrical Y-shaped patterns all pointing in the same direction in any one locality. The convergences are directed down-wind and the regional patterns of the dunes confirms that prevailing wind directions in the past were much the same as they are now. In some parts of Queensland and the Northern Territory dunes of another sort occur. They are reticulate in plan, and their origin is ascribed to the operation of two wind systems of approximately equal strength blowing from different directions.

During the Pleistocene, glaciation on the Australian mainland was of very limited extent. About 50 square kilometres of the highest country around Mt Kosciusko in southern New South Wales was permanently glaciated during the last and most severe glacial phase. As high peaks in the New Guinea mountains were glaciated, it is clear that it was the absence of high mountains in most of mainland Australia which confined the glaciation to such a small area. Snow and ice on other highland areas in the south-eastern part of the country probably came and went with the seasons.

Tasmania, on the other hand, was considerably affected. Recent research suggests that there may have been four major glacial episodes. In the last, about 4000 square kilometres of the north-western part of the central plateau lay under a single ice cap, and there were other smaller ice caps and also valley glaciers. Evidence of glaciation is seen in modern landscapes in Tasmania. Deep lakes occupy glacial valleys, and moraine and erosion features of the mountains tell of the passage of glaciers and of the cycles of ice advance and retreat.

SAND DUNES IN THE GIBSON DESERT

The typical sclerophyll vegetation which characterises most of Australia evolved as the continent dried out. From Gondwanan ancestors there was evolution-in-isolation with the vegetation adapting to the changing environments as the island-continent moved northwards away from Antarctica. The drift away from high latitudes together with the global effects of the ice age profoundly influenced the land and its fauna and flora. Some 40,000 years of human occupancy — and in some places as much as 55,000 years or more — then added another dimension to the factors which were changing the biota. Selection of fire-tolerant plants at the expense of those which could not survive regular burning altered the composition of vegetation over most of the continent.

A Lace Monitor, *Varanus varius*, climbs a tree in search of Birds' eggs. These big Lizards are active predators and scavengers.

JIM FRAZIER

PALAEOGEOGRAPHY OF THE PLEISTOCENE
1 million years ago

Corellas are members of the ancient Gondwanan Parrot family. All the southern continents and India have Parrots and Australia has a great variety. Large flocks of Corellas inhabit semi-arid regions and also the tropical wetlands.

The Fossil Pollen Record in Tasmania also records the fluctuations from glacial to interglacial. During glacial phases the Beech forest contracted and was replaced by alpine grassland. Grasses and ASTERACEAE (Daisy Family) were the main components, and some *Plantago*, *Gentianella* and *Oreomyrrhis* (UMBELLIFERAE) were usually present. Alpine and subalpine heath communities grew in suitable areas and comprised the dwarf Conifers *Diselma* (CUPRESSACEAE), and *Microstrobus* and *Microcachrys* (PODOCARPACEAE), as well as abundant EPACRIDACEAE. During interglacials the forest spread back and changes from one type of vegetation to another appear to have been rapid.

The sea level fluctuations of the Quaternary were often dramatic. Some falls in various parts of the world in the Pleistocene may have been as much as 200 metres. The amount of continental shelf exposed during times of low sea level was obviously very considerable. In Australia, land links with New Guinea and Tasmania have been the rule rather than the exception over the last few million years. Torres Strait to the north did not pose a serious barrier to migration. Bass Strait between Australia and Tasmania, being wider and deeper, has been more of an obstacle. The Bassian landbridge at times of low sea level supported grassland vegetation and areas of scrub heath. The Tasmanian Aborigines walked across it in a glacial and were marooned there when the strait opened and the sea level rose in the last (present) interglacial.

The sea level fluctuations have shaped our coastlines. Terraces are evidence of high sea level stands and drowned coastal valley systems are

EVIDENCE OF GLACIATION IN TASMANIA

MARY WHITE

Deep lakes occupying glacial valleys are characteristic of landscapes which have been subjected to the passage of ice. Dove Lake in Tasmania's central highland region is such a lake. Little Horn and Cradle Mountain tower in the background, their black dolerite cliffs jagged against the sky.

Lakes on the Central Plateau of Tasmania fill hollows in the flat landscape. Ice sheets wore the land down to uniform flatness during the Pleistocene ice age. Dunes of blown sand on one side of some of the larger lakes testify to the dry and windy episodes which characterise glacial times.

MARY WHITE

The Rifle Bird is a Bird Of Paradise — a Rainforest species in which the male has a beautiful and remarkable display routine. He raises his wings in an arch over his head to attract the female to the stump on which he is displaying and during the mating he covers her with his raised wings.

Gouldian Finches are birds of tropical Australia specially adapted for living in open grassland with Pandanus Palms. They are social nesters, several pairs making nests in the same hollow branch or termite mound.

the result of rising seas flooding up valleys which had been deeply gouged by rivers making their way to the sea when it was lower. The pattern of such events is seen in the evolution of Sydney Harbour, Broken Bay and many similar waterways along Australia's coastline.

The amount of land exposed on the continental shelf at times of low sea level was considerable. An example of the sort of areas involved has been disclosed in a recent scientific project carried out by the University Of New South Wales. The aim of the investigation was to determine the ancient drainage patterns on the bedrock below Broken Bay, north of Sydney. Seismic studies showed that at the lowest point, which represents the time of maximum ice cap and lowest sea, sea levels were 125 metres lower than they are today. The deepest bedrock channel is cut down to that level, and when the Hawkesbury River ran through it to enter the ocean it had to travel many kilometres further east than it does today. The contours determined by the seismic studies show that the coastline then was irregular with deep embayments where none exist now.

This bedrock mapping project established that in the case of the shoreline in the Sydney region a broad strip of coastal plain up to 20 kilometres wide extended beyond the present shores at times when ice caps were at their maximum. The land would have emerged gradually as the volume of ice increased, and vegetation would have become established on it progressively. Indeed, an extra strip of coastal plain surrounded the whole continent at those times, wider in parts and narrower in others, depending on the nature of the shelf area. It would have comprised dunes, lagoons, river deltas and sandy areas supporting coastal scrub, heath and grassland.

The continental shelves of all other parts of the world were similarly exposed and submerged as the ice caps fluctuated in size.

The Quaternary Pollen Record shows changes within what is essentially present-day flora and vegetation, whereas the preceding Tertiary record documents the development of plants and vegetation to modern status. The fauna of the Pleistocene part of the Quaternary, in contrast to the familiar vegetation whose components were like those alive now, was still partly "prehistoric". A megafauna of very large animals including giant Kangaroos, Wallabies and Wombats, the huge *Diprotodon*, giant Running Birds and a giant Python were familiar in shape but not in size. Some of them are featured in Aboriginal Dreaming legends, as they co-existed with the earliest Australians and were most probably hunted to extinction by them.

While reasonable analogues for the Quaternary pollen assemblages may be found in modern plant communities, some vegetation types of even 10,000 years ago may no longer exist. In Australia the selection of fire-tolerant species which has occurred in the vegetation since the spread of Aboriginal Australians across the land from 40,000 years ago (and maybe from 60,000 years) has undoubtedly had a profound effect on the structure and content of communities of plants. The concept of the "noble savage" living in complete harmony with the environment is a popular one. There is nowhere better than Australia to counter this myth. The Aborigines were nomadic people without agriculture and with no bronze or metal technology and yet they probably altered the whole aspect of vegetation over much of the continent by their use of fire, in particular as a hunting technique. That they were also a prime cause of the extinction of the megafauna and probably of many less visible and never-recorded animals is also unfortunately true. Similar patterns of early humans causing massive extinctions of the larger animals which lived in America, Europe, Asia and Africa, and the modifications they brought to the vegetation in their enviroments is well documented.

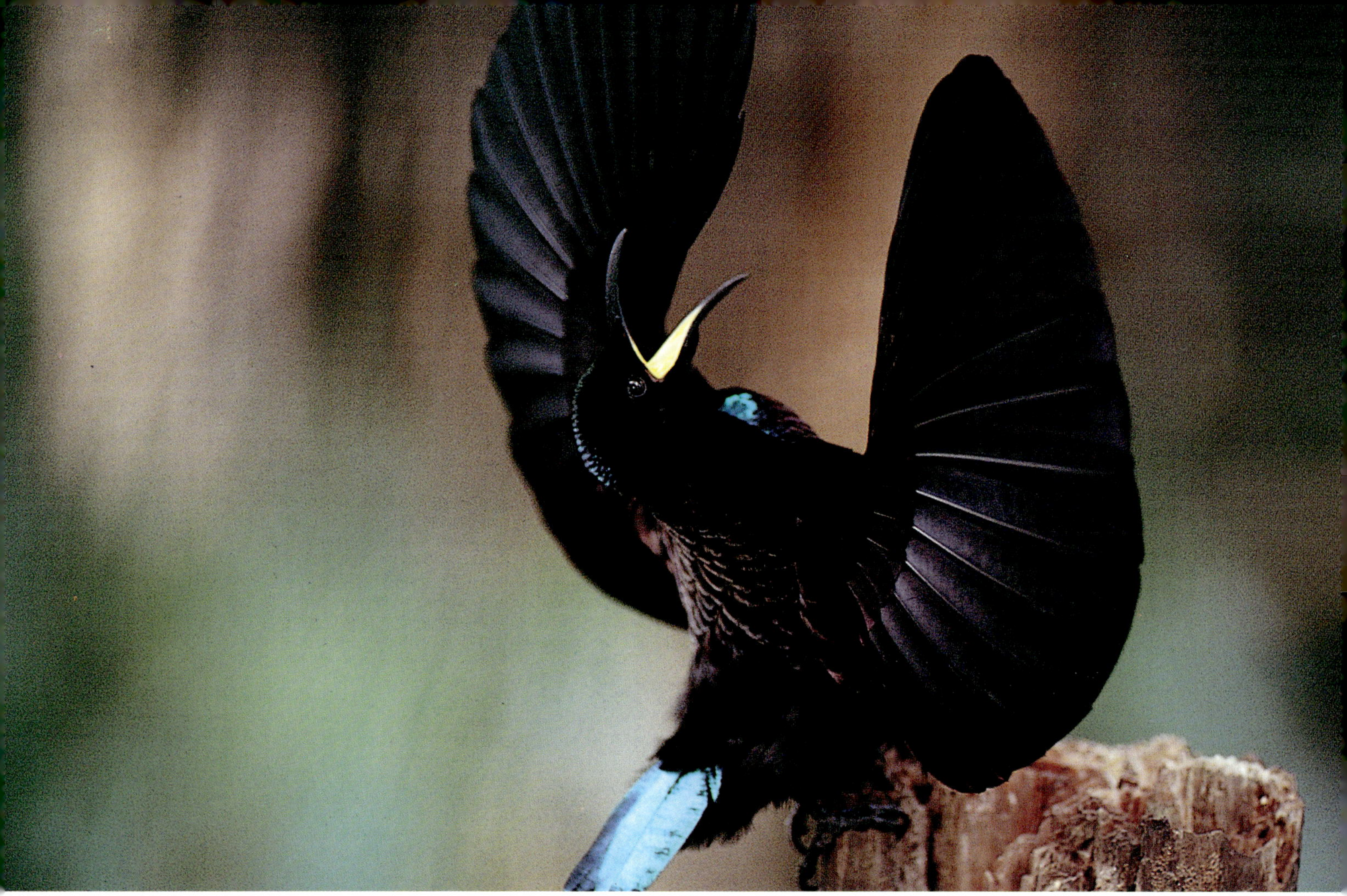

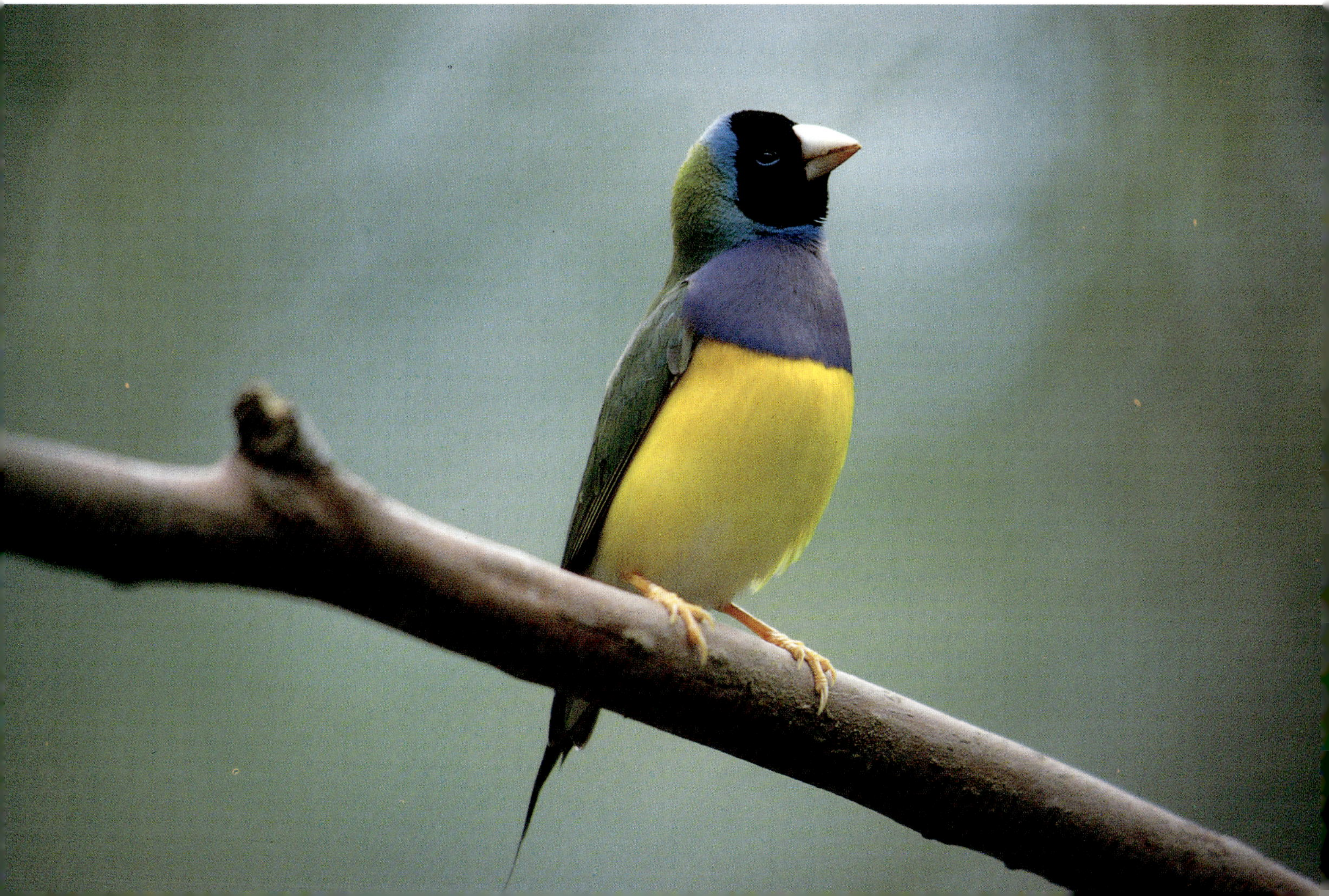

THE QUATERNARY HISTORY OF BROKEN BAY, NEW SOUTH WALES

A brief account of the evolution of Broken Bay, 35 kilometres north of Sydney Harbour, serves as a model for the sort of changes that occurred in coastal regions all around Australia during sea level fluctuations in past ages. It is one of the few waterways in which detailed seismic studies have been carried out, giving us a full picture of the past history of drainage in the area. Modern seismic techniques have enabled the tracing of the old bedrock channels beneath the sediments deposited in the Pleistocene valleys, and also below the sandbars and tombolos along the coast which mask former drainage patterns. (The same sort of scenario as that seen in Broken Bay resulted in the formation of Sydney Harbour.)

Broken Bay is the estuary of the Hawkesbury River. Pittwater, a tidal waterway, opens into Broken Bay between Barrenjoey and Commodore Heights but has a different geological history.

The Hawkesbury is a major river system draining about 22,000 square kilometres of the eastern highlands of New South Wales. The river system, with more or less the same drainage pattern, has existed since Early Tertiary times (about 40 million years). During the Miocene the catchment area was eroded down to form flat plains, and by about 14 million years ago the river system became sluggish and low-energy. A moderate phase of tectonic activity in the Pliocene, about 5 million years ago, uplifted the flat, eroded Miocene plain. As a result the rejuvenated river system cut down through the Triassic rocks, deepening and widening the gorges. Another phase of uplift at the start of the Quaternary repeated the process.

The sea level changes associated with the increases and decreases in the size of the ice caps in the lead up to the Pleistocene ice age, and the fluctuations from glacial to interglacial regimes during the ice age, gave extra power to coastal streams and resulted in further lowering of bedrock in river channels when the sea was low. The old Pleistocene bedrock channel of the Hawkesbury River, between Barrenjoey

and Box Head, is about 125 metres below the present sea level and was the lowest point reached. It is joined in the middle of Broken Bay by tributaries coming from the mouth of Brisbane Water, which was an inland lake system not connected to the sea until The Gap was breached in Recent times. The bedrock bar across The Gap is now only 4.5 metres below sea level and during Pleistocene times it acted as a watershed.

The lowest sea level and the deepest gouging of the bedrock coincided with the widest exposure of the continental shelf. Contours show that when there was maximum exposure about 14,000 years ago, the margin of the land was probably about 20 kilometres further east than it is now. The rivers ran out across a wide coastal plain to reach the sea. The landscapes traversed by the Aboriginal Australians who hunted and fished in the Manly–Warringah area in those times were very different. The ephemeral land must have had lagoons and estuaries, for the rocky bluffs which are now our coastline were far inland. The coastline moved steadily inwards as the sea level rose and was stabilised about 6000 years ago. With the rising sea level, the bedrock channel of the Hawkesbury was progressively silted up, but its large catchment guaranteed its continued active flow across Broken Bay.

The seismic investigation of bedrock drainage patterns in Pittwater contained a surprise, for this waterway which now widens out to merge into Broken Bay did not join the ancient Hawkesbury River drainage channels to enter the sea north of Barrenjoey until very Recent times.

The old drainage pattern etched into the bedrock under Pittwater shows that it was a steep-sided V-shaped valley, parallel to the coast and deepening steadily towards the north. At its watershed in Church Point–Bayview–Mona Vale the valley is shallow and is fed by two tributary valleys, one on the Bayview side, and one from McCarrs Creek. The twin valley systems join abreast of Towlers Bay and a single river channel then winds down to the region of The Basin where a tributary joins in at a depth of about 50 metres. Off Commodore Heights the channel swings abruptly eastward and heads for the ocean across the bedrock between Barrenjoey and Palm Beach where the rock depth is 76 metres.

The failure of the Pittwater River to join the Hawkesbury is due to the presence of a volcanic rock bar which connects Barrenjoey to Commodore Heights and now lies at only 12 metres below sea level. This bar is a Jurassic dyke about 170 million years old. As the seas rose steadily from 14,000

MARY WHITE

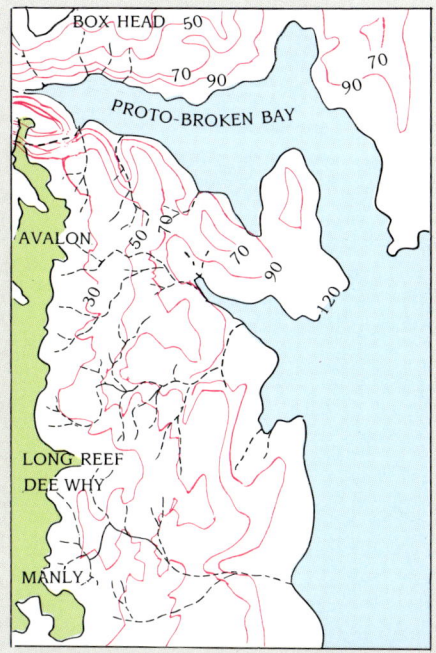

EXPOSURE OF THE INNER CONTINENTAL
SHELF AT THE TIME OF LOWEST
SEA LEVEL
After Albani *et al.*, with permission.

KEY

- present-day land
- land exposed when sea level was 120 m lower than now
- bedrock channels for old drainage
- contours

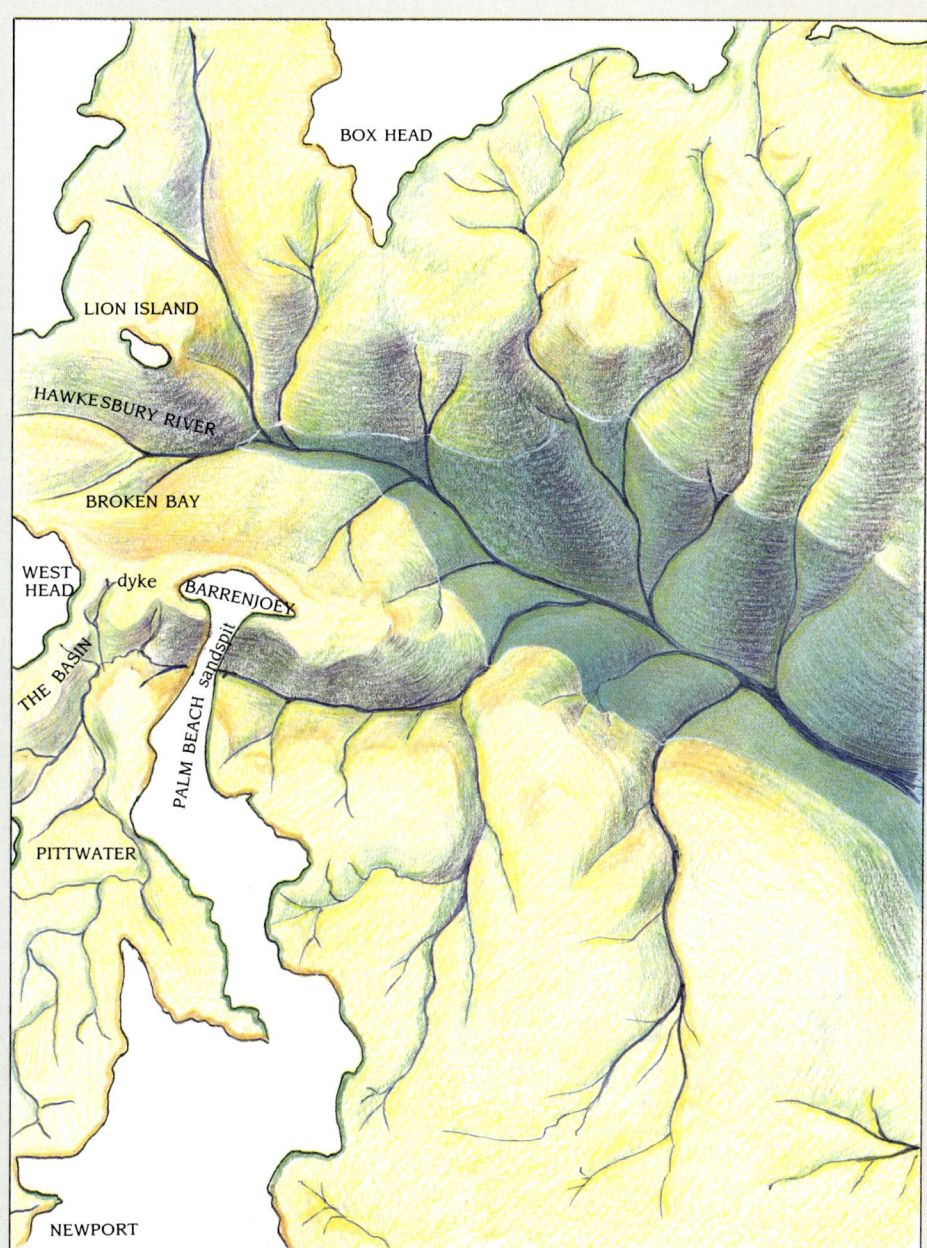

BEDROCK DRAINAGE PATTERNS IN PITTWATER AND BROKEN BAY, NSW
After Albani *et al.* with permission.

years ago, the Pittwater River gradually changed into Pittwater swamp, because, unlike the Hawkesbury, it did not have an active enough flow to counteract the silting of its bedrock channel. Up to 9000 years ago the bedrock from Commodore Heights to Barrenjoey, and beyond seawards, formed a spur of rocky cliff along the southern shore of Broken Bay and the Pittwater drainage was quite separate. Then as the sea rose, the bar was breached and a new drainage pattern was established. The tide started scouring out the sediments which had silted up Pittwater Valley. As the coastal plain narrowed, along-shore currents carrying sand gradually built up the tombolo on the bedrock connection between Barrenjoey and Palm Beach. The old channel where the Pittwater River had run eastwards to

the ocean was silted up and closed. By 6000 years ago the whole system had stabilised. Pittwater had become the tidal system which it is today, wide open to Broken Bay where it joins the flow of the Hawkesbury River and Brisbane Waters.

The rate of sea level rise between 12,000 and 6000 years ago is

calculated to have been 1 metre in 100 years. (The changes to the landscape seen in the study of Broken Bay as the waters rose is a useful yardstick to measure the possible effects of a sea level rise of 1 metre in 20 years which is a worst-case scenario for the Greenhouse Effect, if it is allowed to progress unchecked.)

THE WET TROPICAL RAINFORESTS OF NORTH-EASTERN AUSTRALIA

The relict Rainforests of north-eastern Queensland are unique. They have an unbroken history which goes back 60 or even 80 million years. They are the last remnants of Gondwanan Broad-leaf Rainforest which originated in warm polar regions when the Flowering Plants were spreading across the still-connected southern lands.

Nowhere else in the world is there such a concentration of ancient, primitive Flowering Plants as occurs in these forests. The scientific importance of these last forests is beyond price. Nowhere else on Earth are there forests which contain so much information on the origins, distribution and evolution of Angiosperms, preserved and waiting to be deciphered.

It would be a tragedy for the whole world if the World Heritage listing of these forests fails to protect them from logging and other misuse and violation. The core areas of ancestral forest are closed ecosystems and they require buffer zones of transitional forest types so that they can continue to function as "single living entity" units.

Fan Palms in a Wet Tropical Rainforest in Queensland.

PLEISTOCENE VERTEBRATE FOSSILS FROM NARACOORTE, SOUTH AUSTRALIA

Large numbers of fossils of Pleistocene Vertebrates have been discovered in sediments dated at approximately 35,000 years old. The deposits in which the fossils occur are infilling in limestone caves near Naracoorte, South Australia. The fauna comprises a variety of animals, many now extinct and some still living. Bones of *Diprotodon*, the largest Marsupial ever to have lived, of *Wonambi*, the giant Python, of *Palorchestes*, the giant Marsupial which had a trunked snout, and of many other animals occur as well as the selection of skulls pictured here.

The skull of a Koala, *Phascolarctos cinereus* — the same species as Koalas of the present day. The species is the only surviving representative of a family which has a 25-million-year fossil history. Length of jaw 163 mm.

The skull of a juvenile short-faced browsing Kangaroo, *Sthenurus occidentalis*. Skull length 182 mm.

The skull (jaw length 125 mm) of a Tasmanian Devil, *Sarcophilus laniarus*, which is today the largest surviving carnivorous Marsupial. It was widespread on the Australian mainland prior to the advent of white settlers. Sub-fossil remains dated at 600 years old occur in western Victoria. It is probable that, like the Thylacine (the Tasmanian Tiger), it was ousted from the mainland by the Dingo. Its survival in Tasmania, and that of the Thylacine there until recent times, is explained by the failure of Dingoes to enter Tasmania.

Zaglossus, an extinct long-nosed Echidna. Skull length 179 mm. Members of the genus still exist in New Guinea Rainforests.

Thylacoleo carnifex, a Marsupial Lion. Skull length 290 mm.

NEW ZEALAND IN THE QUATERNARY PERIOD

The Quaternary geological history of New Zealand is one of great unrest and change. It encompasses dramatic tectonic events in which mountains have continued to be thrust up along the Alpine Fault, and, though rapidly eroded, they remain alpine in their magnitude.

In the Pleistocene Epoch, widespread volcanic activity was locally devastating. The sea encroached onto the edges of the land and also flooded wide areas to form temporary marine basins, before retreating and leaving wide continental shelf exposures. There were other basin areas, created by tectonic movements, where freshwater sediments, brought by rivers from the fast eroding mountains, accumulated. It was enough change, one would think, for the fauna and flora to cope with, without the added imposition of ice age regimes.

The ice age had a far more profound effect on New Zealand than it did on Australia, where the effect was moderated by the large size of the continent, the situation of much of its land in lower latitudes and the absence of high mountains. Surprisingly some of the Gondwana-derived flora survived all the exigencies of physical and climatic change in New Zealand. We can learn a great deal about expansion and contraction of vegetation types and the significance of refugia, however small, by seeing how reforestation and regeneration of vegetation has taken place in New Zealand in the recuperative period since the last glacial event about 14,000 years ago.

Paroxysms of the Kaikoura Orogeny resulted in uplift and deformation of rocks on a scale almost beyond comprehension. It has been estimated that vertical uplift to the east of the Alpine Fault in central Westland was in the order of 18,000 metres, with the horizontal movement being of approximately the same order. The amount of erosion since the Mid Pleistocene (in about the last million years) can be calculated to have removed 15,000 metres of height from the range, leaving Mt Cook at 3763 metres as a mere remnant. Somewhat less dramatic activity was occurring along the whole Alpine Fault zone. The climax of the orogeny was accompanied by much surface volcanism, the effects of which can be seen in many places. Volcanic activity on a reduced scale continues right up to the present, as many of the major structures and associated fractures that developed in the Pleistocene are still active.

Early in the Pleistocene Epoch the volcanic activity of the Northland Peninsula continued in the Bay Of Islands and Whangarei, but was declining in intensity as it migrated southwards to become localised in the Central Volcanic Belt. Late Pleistocene eruptions (both effusive and explosive) took place in the present-day Auckland city area until as recently as about the year 1200 AD. In the south Auckland region the prominent eroded cones of Pirongia and Karioi are remnants of volcanoes which were erupting violently from the Late Pliocene to the Early Pleistocene. The volcanic rocks of Little Barrier Island and of Major Island are probably of Mid Pleistocene age.

The principal igneous activity of Pleistocene times was concentrated in the Central Volcanic Zone. Here ignimbrites were produced in massive sheets as melted crust gushed out from the Taupo Volcanic Zone, making space for the subsiding of the Taupo graben (a basin structure bounded by faults). The ignimbrite sheets are sub-horizontal and form two plateaux sloping gently north-north-east from Tongariro National Park to the Bay Of Plenty on either side of the Taupo subsidence zone. Edges of the sheets form conspicuous escarpments in the landscape. The air-fall pumice produced by the eruption of the ignimbrites forms a layer on the seafloor of

A New Zealand Kauri (Agathis) forest. Agathis is a Southern Conifer, an Araucarian. While small areas of pure Kauri stands still exist in New Zealand, Australia has lost most of this Gondwanan forest type.

GEORGE POINAR

the adjacent east coastline, and beach sands are often largely pumice. The immediate surrounds of Lake Taupo are all pumice sands.

The Central Volcanic Region complex of five major volcanoes is an awesome sight today. Its surrounding areas of mudslips and its desert-like landscapes show us how many areas were similarly devastated in the not-so-distant past. The late stage thermal and sulphur vapour phase of the Rotorua and Taupo volcanics is now the most important economic resource of the district. Deep drilling to obtain access to geothermal steam allows steam-driven power stations to be developed.

In western Taranaki, volcanism has occurred since Miocene times, apparently with a southwards progression of vents along a line which ended in Mt Egmont (Taranaki) around 1500 AD.

South Island, in contrast to North Island, had little volcanism in the Quaternary. On the Banks Peninsula, two Pliocene volcanoes were still active in the Early and Mid Pleistocene, and then died. Their craters, breached by the sea, are used as harbours. Another on the Otago Peninsula has a similar history.

The sea level fluctuations which resulted from growth and contraction of ice caps affected coastal environments. Variations of up to 150 metres are suggested for New Zealand. The widened coastal plains at times of low sea level resulted in rivers cutting deep gorges to run into the sea. Many coastal landscapes clearly show the changing sea levels. In Wellington Harbour there are eight old shorelines forming wide bench-like terraces ranging from 25 metres to 370 metres above the present sea level. (The continued elevation of the Rimutaka Range during the Quaternary is responsible for the raising of some of these terraces.)

Glaciations seriously affected South Island because of the height and extent of its mountain ranges. The snowline was depressed by about 1000 metres (compared with the present one) and there was a reduction of about

Mt Cook, at 3764 metres, is the highest peak in New Zealand. It is located in the alpine mountains of South Island.

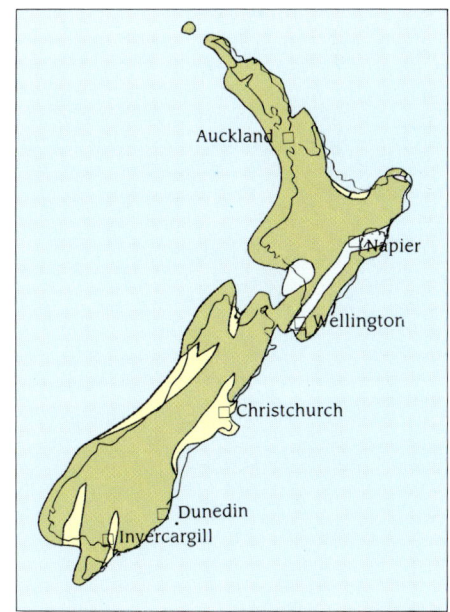

PALAEOGEOGRAPHY OF NEW ZEALAND IN THE EARLY QUATERNARY

231

6°C in mean temperature. Glaciers extended along the main alpine axis, and most of the western coast zone was frozen as well, under the associated snowfields. Only small areas in Banks Peninsula, the north Canterbury coast, north Marlborough and north-west Nelson had summer temperatures above 10°C. Most of the present lowland area and its ephemeral coastal extension were part of the periglacial zone. In the Westland and Canterbury regions six or seven distinct glacial advances can be recognised.

The geological effects of the ice age regimes are manifest in South Island. They include deep erosion in the rising mountains and the formation of moraines. The deep fiords in the south-western region were caused by glaciers, and many inland glacial valleys now contain deep lakes. Outwash sediments from the glacial fronts on the eastern side of the mountains formed great alluvial fans, spreading out to form the Canterbury Plain. Large parts of the eastern lowlands are covered by wind-blown "loess". This unconsolidated fine sediment is rock flour produced by the action of glaciers grinding across a landscape, which has since been blown into dunes.

During the ice age fluctuations, vegetation changes were drastic in South Island. A shifting mosaic of different vegetation types was moving in accordance with the changing circumstances in those parts which had not been made inhospitable by snow and ice. Coastal fringes, perhaps the marginal areas created by low sea level in particular, would have been a refuge for an assortment of plants allowing them to survive and be available to spread back into suitable areas as conditions improved. The ice-affected areas were deforested and in interglacials were subject to rapid erosion. The present forests in South Island are the result of natural reforestation since the last glacial 14,000 years ago. During the last glacial phase the only extensive forests in New Zealand were in North Island.

In North Island, glaciers were formed only on the high peaks of the Central Volcanic Massif and on Taranaki. A zone around the mountains may have been tundra-like with summer temperatures below 10°C, but most of the island would have been able to support mixed forest. Sea level changes affected the island at the start of Quaternary time. In the first warm interglacial, many areas which had been emergent in the Pliocene were temporarily re-flooded and the Manawatu Strait again connected marine basins to its east and west.

Rocks of the Late Pleistocene in North Island, especially in the west, contain the products of the Central Zone volcanoes. Coastal terrace deposits

THE TUATARA

The Tuatara (Sphenodon punctatus), New Zealand's "Dragon", is possibly a direct descendant of the Dinosaurs. It is an engaging animal, so why do our reconstructions of its ancestors always portray them as formidable and without charm? Simply because Dinosaurs were large and prehistoric need not mean that they were any less elegant than the scaled-down living Tuatara.

New research, using sophisticated techniques of DNA matching, is questioning the ancestry of the Tuatara and suggesting that it may be more closely related to Snake and Lizard ancestral lines than to Dinosaurs. At present, though, it is considered of Dinosaur ancestry by most authorities.

An adult Tuatara.

HAL COGGER

The "Weta" occupies the small Rodent niche on the forest floor. Until humans brought Rats to New Zealand there were no small Mammals other than Bats, so large Insects were able to fulfil the role of scavengers among the leaf litter.

mark the stands of high sea level during interglacials, and river-borne sediments characterise glacials when erosion was rapid because rivers were cutting down deep to reach the sea. Loess deposits in central North Island are the result of glaciers grinding the rocks to rock flour as they moved down the mountains.

The recurrent glacial episodes in New Zealand markedly changed the vegetation, and warm-temperate forms disappeared from the Pollen Record. Transitional pollen floras from the Tertiary–Quaternary boundary in the south Auckland region of North Island give an idea of the fluctuations in the vegetation which were occurring in response to the wetter/drier, warmer/colder changes of the ice age regimes. A flora at Huntly shows changes from *Nothofagus brassii* dominance to *Nothofagus fusca* in the Late Miocene to the Pliocene, and from *Nothofagus fusca* to Podocarp Conifer dominance in the Pleistocene. At Frankton, in assemblages in which an abundance of RESTIONACEAE indicates marshy conditions, there is a sequence from forests dominated by *Nothofagus brassii* and *Nothofagus fusca* in equal proportions (interglacial) to cooler forests with a high proportion of Podocarps including the Celery-top Pine (glacial). Then follows a resurgence in the proportion of *Nothofagus brassii* pollen indicating another interglacial, followed by another cooling episode with *Nothofagus fusca* Beech, *Dacrydium cupressinum* (Rimu) and *Phyllocladus* (Celery-top Pine) dominating. Cooler intervals are clearly shown by the *Phyllocladus* curve when a graph is made plotting its pollen frequency against temperature.

In a locality on the Oraurangi Peninsula in west North Island, pollen of MYRTACEAE and *Nothofagus brassii* dominate in an interesting assemblage which includes: *Metrosideros* (Rata), *Eucalyptus*-like pollen and *Leptospermum* (Tea Tree), all of MYRTACEAE; *Nothofagus menziesii* Beech; PROTEACEAE; CHLORANTHACEAE; RUBIACEAE; *Dodonea viscosa* (SAPINDACEAE); *Acacia* (MIMOSACEAE); and Podocarp Conifers (Kahikatea). This community suggests that conditions locally were fairly dry and possibly windy.

As the ice age proceeded, a number of taxa became extinct in New Zealand. PROTEACEAE were reduced to two species, *Knightia excelsa* (Rewarewa) and *Persoonia toru* (Toru). *Acacia* disappeared. But the variety of habitats and environments, and the presence of refugia for species temporarily excluded from areas by unsuitable conditions, enabled the Modern New Zealand Flora to survive and regroup to form the modern vegetation patterns after the last glacial period.

The marine Invertebrate Fossil Record for the Pleistocene shows changes in composition according to the fluctuations from glacial to interglacial, and many warm water forms disappear. The Vertebrate Fossil Record documents the racial history of the Moas and other flightless Birds. There is no record of Flying Birds until the Late Pleistocene, when a flightless Eagle (which presumably lost the power of flight because it was no longer needed) lived in the forests. Modern Penguins also appear in the Fossil Record in the Late Pleistocene.

In view of New Zealand's turbulent history, with the ephemeral nature of much of its land through time and all the environmental stresses to which plants and animals were subjected, it is surprising that so much of its Gondwanan inheritance of living things survives into the present. But it is no surprise that the Vertebrate fauna is so impoverished or that the flora has so few native species when compared with that of other lands of similar size elsewhere in the world.

From a scientific point of view the importance of New Zealand's Cainozoic floras cannot be overestimated, because of their wholly Cretaceous ancestry and their evolution-in-isolation since.

Nothofagus forests in South Island.

GEORGE POINAR

Aciphylla sp., the Flowering Spaniard, a
member of the APIACEAE, the family to which
the Flannel Flower (*Actinotus*) also
belongs. Both genera are restricted to
Australia and New Zealand, while the
family has cosmopolitan distribution.

GEORGE POINAR

Beech forest (*Nothofagus*) habitat,
South Island.

*Remains of an Agathis (Kauri) forest
preserved in a swamp. The tree trunks are
dated at 30,000 years.*

GEORGE POINAR

The trunk of a Kauri Pine covered with epiphytes. Epiphytes are a feature of New Zealand forests, and trees are often festooned with an assortment of different kinds from their bases right up into their crowns.

Agathis forest habitat, North Island, New Zealand.

JOHN FORLONGE

An Araucarian Conifer emergent from Nothofagus woodland in Chile.

NEW ZEALAND, LAND OF BIRDS

When the first humans — Polynesians from the Pacific Islands — took up residence in New Zealand 1000 years ago, they represented the Placental Mammal invasion of a land which up to that time might have been the most exclusively Bird-dominated land of its size ever to have existed. As a consequence of its history — the time at which it severed its connections with Gondwana, and the complement of its Gondwanan biota on separation — it was unique. If New Zealand ever had early Marsupials and Monotremes they did not survive, and they have not been found in the Fossil Record. Its largest Reptile, the Tuatara, which may be evolved from Dinosaur stock (or else from the Lizard and Snake line of equal age) has survived because of New Zealand's isolation and the lack of competition.

At the time that the earliest Maoris arrived the land was largely forested. The fauna was impoverished because of the exigencies of the environmental changes through the ages and the absence of supplementation by migrants because of extreme isolation. Marine Mammals, notably Seals, Dolphins and Whales, lived in the surrounding ocean. On land there were: Insects, including the strange ''Weta'' which occupies the small Rodent niche on the forest floor; other Invertebrates; a few Reptiles and Frogs; two species of Bat, which at that time were the only land Mammals; and the all-important Birds. The Gondwanan Running Bird descendants had diversified to fill all the niches which in other lands are occupied by Mammals. One evolutionary line led to the grub-eating Kiwis, which still survive, and the other to the now-extinct Moas. The largest Moas, standing to 3 metres high, correspond to large herbivores of other forested lands; the medium-sized were the ''Antelope equivalent'', and the smaller ones were the ''Rabbits'' or other small herbivore equivalents. When the Maoris arrived, the forests were networked by Moa pathways, well worn by the thousands of Birds which lived and grazed in them.

The forest cover was more or less intact until about 700 years ago. By about 500 years ago Maori activities had resulted in the complete loss, as a result of burning, of forests on the eastern coast of South Island and of the lowland forest on the western coast of North Island. Forest had also contracted in other areas.

For the Maoris, the Moas (whose name means ''common fowl'') were a source of food. Carbon-dating of Moa bones at Maori campsites confirms that hunting started 1000 years ago and continued for at least 500 years, and by compiling information on the dates of sites all over New Zealand a picture emerges of the sequence of events. It shows that the hunters moved camp when they had used up the Moa resource in an area. North Island was hunted out first, and when there were no more Moas there, South Island was worked, with the youngest sites being in the extreme south. The complete documentation of the demise of this large and important group of animals so clearly shows humans to be the agents causing the extinction, over a short period of time and in spite of their own relatively low population numbers. Moas were not only a source of food. Their bones were used to fashion fish hooks, harpoon heads and ornaments, and their feathers and skins were incorporated into clothing. Eggs were used as water containers.

Because of the absence of predators in New Zealand, many other normally flighted kinds of Birds lost the power of flight over time. A large Goose, a giant Woodhen or Rail, and a Duck were among flightless forms which were unable to escape from human hunters, and they suffered the same fate as the Moas. A very large Eagle which probably had only limited powers of flight also became extinct: its prey was Moas, and when they ceased to exist, it did too.

THE EXTINCT GIANT EAGLE
AFTER DON BRATHWAITE, WITH PERMISSION

237

DIVARICATING PLANTS AND MOA BROWSING

The New Zealand flora contains an unusually high percentage of small-leaved, woody shrubs with closely interlaced branches. They are described as "divaricating", which indicates branching at a wide angle. The 53 species of this flora belong to 17 families and a larger number of genera, and most of them grow on fertile soils. All tend to have tough stems, and the leaves on their outer branches are smaller and more widely spaced than those within the clustered branches.

The New Zealand divaricating plants are not spiny, with one notable exception, the Matagouri (*Discaria tuomatou*), which is abundant in South Island. Those species which grow as shrubs maintain the divaricating habit throughout their lives. Some others, which have an adult tree-form, are divaricating in the juvenile stage only.

It has been suggested that the peculiar habit of divaricating plants is the result of adaptation to prevent damage by browsing Moas. After being grazed, plants would have resprouted. Shoots forming at wide angles, particularly those which grew inwards to the centre of the plant, would have been better protected against the clamping, pulling and breaking feeding-action of the strong-beaked Moas. Natural Selection would have favoured the kind of plants whose structure minimised damage by grazing animals.

Moas comprised six or seven genera and a dozen or more species. They were abundant, and most were adapted for forest living. The Moa had evolved from a Gondwanan Running Bird ancestor, and the Kiwi is a surviving member of another line of evolution from that same ancestor. The other southern lands which were united in Gondwana also have big, flightless birds descended from this same Gondwanan ancestor: South America has its Rheas; Africa has Ostriches; Madagascar had the now-extinct Elephant Birds; and Australia had the now-extinct Dromornithids and still has the Emu and Cassowary.

In some localised dry and open habitats in other parts of the world, the divaricating habit is also seen, though

MICHAEL GREENWOOD

Elaeocarpus hookeriana, a divaricating plant. This specimen shows the complicated branching which is a feature only of juveniles in this species.

the plants are all spiny and there are far fewer species than in New Zealand. In Patagonia, Madagascar and California, divaricating plants have probably evolved in response to grazing by Mammals. Their spiny nature would act as a deterrent to soft-nosed browsers. In New Zealand there was no advantage in having spines as the hard, bony beaks of Moas were less sensitive to such protective mechanisms.

When Maori people arrived in New Zealand 1000 years ago, Moas were abundant. Running Birds, up to that time, had been safe as there were no predators. But they were no match for human hunters and egg-gatherers, nor for the Dogs and Rats which came with the Maoris. Destruction of the Moas' habitats by fire was another factor which ultimately led to their extinction. Moas had been long gone before European settlers arrived in the land. As well as the abundant fossil bones and egg shells of Moas found in New Zealand, there is much material which is not old enough to be "fossil". Examples of this sort include mummified remains which supply evidence of feathers, gut content (comprising twigs and grass) and even skin samples suitable for analysis of DNA. In some areas Moa bones were so abundant that they posed problems for farmers ploughing the land, and cartloads of bones were sent away to bone mills to be ground into fertiliser.

MOA

AFTER DON BRATHWAITE, WITH PERMISSION

JIM FRAZIER

The Plumed Egret, Egretta intermedia. *A symbol of life in a green world where all life owes its being to the green colour of the Plant Kingdom. The Nature of hidden worlds of the past, and the nature of our world of the present, was predestined nearly four billion years ago by a green revolution . . . and re-greening is the cure for the man-made sickness that is slowly and surely killing our planet. Where there is green there is life, and the seeds of hope are borne by plants. Earth's history, written in the rocks, tells us that Man is but a temporary custodian, insignificant in the overall scheme of things . . . .*

EPILOGUE

The story of the changing world in which prehistoric animals lived through the long ages of geological time is one of changing environments which caused extinctions and which promoted the evolution of new forms. The changes were "natural" changes, with natural laws keeping the balances, and as a result the Earth was a healthy living entity with the interactions between all its systems, animate and inanimate, in harmony.

Then came *Homo sapiens*, a species which like a disease affects the health of the Earth and threatens its very life. The Fossil Record of the human species on Earth is very short — only a brief record of its earliest steps towards cultural and technological evolution. When time has moved on a few more million years the record may well contain a "Terminal Holocene Event", perhaps marked by a plastic layer worldwide, in the manner of the iridium layer which millions of years ago indicated another crucial time of global stress and change. And *Homo sapiens* will be seen as the not-so-wise Ape which broke all the rules and killed the Earth that had given him life.

Removal of big logs from this forest near Laurieton in New South Wales has vandalised the rest — broken trees, shocking waste and desecration typify current careless attitudes.

MARY WHITE

It may not be too late to prevent such a catastrophe. Maybe with a concerted effort, the sum of small efforts made by everyone on Earth, we can cure the sickness and restore the Earth to health.

In Australia we must ask the question: what have we done with the legacy we inherited when we, the human race, came to the pristine island-continent?

The answer has two parts, one dating back perhaps 60,000 years and the other only 200 years.

The Aboriginal Australians were not to know that their firing of the land was changing the composition of the vegetation or that their hunting was altering the fauna to the degree that many of the animals they pursued would soon exist only in their Dreamtime legends. Because they were not making sudden, drastic changes to the environment the Aborigines' presence equated to the rise of animal species in the past: their arrival on the scene resulted in a decline of some of the species with which they competed, and was thus a "natural" process such as has formed the basis of evolution and change in the biota since the beginning.

But that was then, and this is now. . .

While natural processes of change characterised the thousands of years of Aboriginal occupancy, the changes wrought by 200 years of settlement by Technological Man are mostly "unnatural" — too great, too rapid, too blindly oblivious of the consequences to the biota. Many of these changes were started in ignorance, but they are now continued in spite of the knowledge of their consequences. The problem exists largely because political and financial forces govern, where head and heart should dictate the rules. When our forests are gone, our soils are eroded to dust, our weather patterns have changed, and survival becomes a losing battle for most creatures, it will be too late to stop the slide towards extinction.

It is not acceptable, for example: that the last of our tall forests are threatened and our remaining wildernesses are being violated; that Duck shooters are licensed to kill embattled, beautiful creatures whose habitats are disappearing, and hunters are allowed to kill other creatures for "sport"; that special rules apply to Aborigines allowing them to kill the Turtles and Marsupials and Birds which are protected from the predations of others, when they no longer rely on them for food; that remaining wetlands be drained to allow expansion of cities; and above all that the sea is used as a sewer and we are killing it slowly and surely. The list of dangerous practices which are part of the accepted pattern of modern life, regardless of their consequences to the environment and their disastrous effects on the balances of the natural world, is endless.

The answer to the question "what have we done. . .?" becomes another question: what are you, and I, and everyone on this Earth going to do to turn the tide and ensure that there will be a future for our children, and that it will be a bright one in a bright and beautiful world?

GLOSSARY

abyssal current The current which flows on the bottom of the deep oceans, moving very cold water from the South Pole northwards.

acellular organisms Protists, single-celled organisms in which all the functions of a many-celled organism are carried out by the one cell.

Algae A phylum of primitive plants, mainly aquatic, without conducting tissue. Includes seaweeds and marine phytoplankton.

alluvium Unconsolidated sedimentary material which was carried in suspension by rivers and deposited on floodplains, deltas or the river beds. Silts, sands and gravels, which may form fertile alluvial soil.

alternation of generations Life cycles with a repeated pattern of asexual and sexual reproduction. A sporophyte plant with double chromosome numbers produces spores with half the chromosome number by a type of division known as meiosis. Spores produce haploid gametophytes which produce male and female gametes. Fusion of gametes produces a diploid zygote, and from this the new sporophyte plant develops.

Ammonites Molluscs with coiled, partitioned shells with convoluted sutures. Range: Devonian to Cretaceous.

Amphibians Cold-blooded Vertebrates which breed and pass through the larval stages of development in the water, but spend their adult stages primarily on land.

Ancestral New Zealand landmass The large area of land which arose from the sea in the Late Jurassic, and extended from the Campbell Plateau off Antarctica to New Caledonia. Modern New Zealand is a small surviving relic of this ancestral land.

andesite Fine-grained igneous rock, generated at subduction zones, which forms new continental crust.

Angiosperms The Flowering Plants — vascular land-plants with seeds contained in ovaries. They evolved in the Cretaceous and dominate world flora today.

Animal Kingdom Living creatures, typically mobile, unable to make their own food — unlike the members of the Plant Kingdom which do so by photosynthesis.

Annelids Phylum of soft-bodied creatures — Worms and Leeches.

Archaeocyathids Early coralline organisms which are ancestral Sponges. Abundant in the Cambrian, then replaced by conventional Sponges.

Arthropods Members of the phylum Arthropoda, having jointed limbs, a segmented body and an exoskeleton.

autotrophic organisms Those organisms capable of making their own food within their cells.

aquifer A pervious or porous layer of rock which is saturated with water and can feed springs or wells.

asexual reproduction Reproduction without eggs and sperm.

Bacteria Kingdom of microscopic organisms.

basalt Fine-grained igneous rock. Principally occurs in lava flows and constitutes over 90 per cent of volcanic rocks.

basin A hollow, downwarp or area of continued subsidence in the Earth's crust, in which sediments accumulate over a long period.

batholith A very large mass of intrusive rock, usually granite.

Belemnites Extinct marine Molluscs (Cephalopods) with bullet-shaped cases. Range: Mesozoic.

benthic organisms Those that live on the bottom of the sea.

biosphere The regions of the Earth in which life exists.

biota The plants and animals which occupy a specific region and/or period of time.

Bivalves Pelecypods — Molluscs with two shells. Range: Cambrian to Recent.

Brachiopods Lamp Shells — marine Invertebrates of the phylum Brachiopoda, with two shells. They look like Bivalve Molluscs but have different anatomy and symmetry. Range: Cambrian to Recent.

braid-plain A plain over which rivers meander and anastomose.

Bryozoans Colonial animals (Moss Animals) which attach to rocks, other animals or the seafloor. Range: Ordovician to Recent.

Cainozoic Era The most recent Era, comprising the Tertiary and Quaternary Period. The last 65 million years.

calcareous Containing or consisting of calcium carbonate.

Cambrian Period The first Period of the Phanerozoic Eon. From 570 to 505 million years ago.

carbonaceous Containing carbon, especially sedimentary rocks containing organic matter.

carbonates A group of widespread minerals, primary sedimentary constituents.

Capitosaurs An order of Reptiles.

Carboniferous Period A Palaeozoic Period from 360 to 286 million years ago, between the Devonian and the Permian Periods.

cellulose The substance comprising the cell walls in plants.

Cephalopod A marine Mollusc of the class which includes Squid, Octopi and Ammonites. Range: Cambrian to Present.

chitin A horny substance forming the hard cover of Crustaceans, Insects, and others.

chlorophyll The green pigment essential for food manufacture in plants.

chloroplast An organelle in which chlorophyll is contained.

chromosome Thread-like structure which carries genes in the nucleus.

Chordates Animals with spinal columns.

cilia Protoplasmic extensions of a cell which beat and propel the organism.

class Subdivision of a phylum.

climate The sum of long-term weather conditions, determined by the location of an area, its latitude, altitude and situation in relation to the main circulation belts of atmosphere and ocean.

Closed Forest A self-sustaining ecosystem which recycles its nutrients.

Clubmoss A member of the Lycopod phylum of spore-bearing plants.

coccoliths Microscopic calcareous

plates borne by calcareous unicellular Algae. Can accumulate on the ocean floor in astronomical numbers and form limestone. Since the Jurassic they have formed "nannoplankton".

Coelenterates   Members of a phylum of Invertebrates which includes Jellyfish, Sea Anemones and Coral. Range: Proterozoic to Present.

conglomerate   A coarse-grained sedimentary rock formed largely of rounded, water-worn pebbles.

Conifer   Cone-bearing, gymnospermous plant.

Conodontophorids   Animals which had Conodonts as their filtering or eating mechanisms.

Conodonts   Microscopic phosphatic tooth-plates. Range: Cambrian to Triassic.

Continental Drift   The relative movement of continents over the Earth's surface as a result of seafloor spreading.

Coral   Coelenterates whose hard, calcareous skeletons form reefs in tropical waters. Range: Ordovician to Recent.

Cordaites   An order of early Gymnosperms which existed in the Palaeozoic Era and are assumed to be ancestral to most modern Conifers. Range: Devonian to Permian.

Coriolis Force   A force caused by the spinning of the Earth, deflecting currents to the right in the Northern Hemisphere and to the left in the Southern Hemisphere.

Cretaceous Period   The last Period in the Mesozoic Era, from 144 to 66.4 million years ago, between the Jurassic and the Tertiary Periods.

Crustacea   A Class of marine Arthropods.

Cryptozoic Eon   The Age Of Hidden Life — the 4000 million years of Earth history before the Cambrian Period.

cuticle   The waxy outer layer on the surface of plants which reduces water loss.

Cyanobacteria   Primitive organisms which are able to photosynthesise. The oldest fossils are those produced by these organisms — Stromatolites, dating from 3500 million years ago and with descendants living today.

Cycadophytes   Phylum of gymnospermous plants to which several extinct orders and modern Cycads belong. Range: Late Permian to Recent.

cytoplasm   The protoplasm of cells outside the nucleus.

delta   A generally fan-shaped accumulation of river sediment at the mouth of a river.

Devonian Period   The Period between the Silurian and the Carboniferous, in the Palaeozoic Era. From 408 to 360 million years ago.

Diatoms   Microscopic unicellular Algae with siliceous tests, abundant in plankton.

diatomaceous earth   Pure siliceous rock composed of the shells of Diatoms.

diatreme   A volcanic pile or vent.

Dicotyledons   Class of Flowering Plants in which embryos have two seed-leaves.

Dinoflagellates   Unicellular planktonic Algae. Range: Mesozoic to Recent.

Dinosaur   Extinct terrestrial Reptile. Range: Mesozoic.

diploid   Organism or cell with a double chromosome number.

division   A phylum in the Plant Kingdom classification is often referred to as a division.

dolerite   Igneous rock of minor intrusions.

dolomite   A type of limestone.

dropstone   A boulder which was floated by ice and dropped into sediments at some distance from its source.

dyke   A vertical intrusion of igneous rock which comes up along a fault or a line of weakness and cuts through the pre-existing rock of an area.

Echinoderms   Members of a phylum of Invertebrates which includes Star-fish, Sea Cucumbers, Sea Urchins and Crinoids.

ecosystem   A specific habitat and its living organisms which behave as an integrated unit.

egg   A female gamete.

embryo   The structure formed after an egg has been fertilised. It develops into a new individual.

endemics   Organisms which occur only within a specific region.

endosperm   A store of food in seeds of Angiosperms which nourishes the embryo.

environment   That complex entity comprising all the conditions, circumstances and influences which surround and affect the development of an organism or a group of organisms.

Eocene Epoch   Subdivision of the Tertiary Period. From 58 to 36 million years ago.

Eon   Major subdivision of geological time.

Epoch   Subdivision of a Period.

Era   The Phanerozoic Eon is divided into the Palaeozoic, Mesozoic and Cainozoic Eras.

erosion   Lowering of the land surface by weathering, corrosion and transportation, under the influence of gravity, wind and running water.

Eukaryotes   Organisms whose cells contain well-defined organelles, including a nucleus.

Eurypterids   Scorpion-like Arthropods. Range: Ordovician to Permian.

evaporites   Any sedimentary rock formed by precipitation from saline water.

exoskeleton   The hard outer structure protecting and supporting the body of certain animals, for example, Arthropods.

extinction   The disappearance of all members of a species, genus or family of organisms.

exine   The outer coat of pollen grains.

family   In classification, a grouping of genera with similar characters.

fault   A fracture along which rocks on one side are displaced relatively to those on the other side.

fauna   The animal population.

Fern   A member of the Filicales — spore-bearing vascular plants which have an alternation of generations in their life cycle.

flagella   A whip-like protoplasmic projection which propels a mobile organism.

flower   A structure composed of specialised leaves (sepals and petals)

243

together with the reproductive structures of stamens (male) and ovary, style and stigma (female) in Angiosperms.

Flowering Plants Angiosperms, in which seeds are enclosed in an ovary.

Foraminifera Order of marine Protistan organisms with calcareous shells. Accumulations of the microscopic shells have contributed to limestones since Devonian times. From the Cambrian to the Jurassic these organisms were benthic, since then they are abundant in plankton. Range: Cambrian to Recent.

fossils Evidence of life preserved in rocks.

Fossil Record The sequence of fossils through geological time.

fruit A ripened ovary.

Fungi Kingdom of thallophytes, which lack chlorophyll and therefore do not have the power to photosynthesise. Saprophytic, parasitic and symbiotic.

Gaia The Earth Mother — the concept of the Earth acting like a single living organism.

gamete A sperm or egg-cell prior to fertilisation.

gametophyte A plant which produces sex cells (gametes). Ferns and others which have an alternation of generations have both sporophyte (spore-producing) and gametophyte (gamete producing) plants.

Gastropods Molluscs with coiled shells. Range: Cambrian to Recent.

gene The unit of inheritance, carried on chromosomes in the nucleus of cells.

genus A unit of classification into which species are grouped.

geological time The 4600 million years since the Earth was formed.

Ginkgophytes Phylum of gymnospermous plants with a long fossil history, but with only one living species remaining today. Range: Permian to Recent.

glacier A river of ice.

Globigerina **ooze** Deep sea ooze (blanket of fine sediment on the ocean floor composed of microscopic tests of Foraminiferids).

Gondwana The great southern supercontinent.

graben An elongated body of rock down-faulted between parallel normal faults.

Graptolites An order of marine colonial organisms (free-swimming) with a spinal canal but no backbone, which were related to Vertebrates and members of the Chordata. Range: Cambrian to Carboniferous.

Greenhouse Effect The warming up of the Earth as the result of an increase in carbon dioxide in the atmosphere, and all the problems attendant on it.

Gymnosperms Seed Plants whose seeds are not enclosed in an ovary, for example, Conifers, Cycads, Ginkgophytes and Seed-ferns.

gypsum Calcium sulphate, an evaporite.

heterotrophic Bacteria Bacteria incapable of synthesising their own food.

Horseshoe Crabs Arthropods with large head shields. Only one genus survives today. Range: Ordovician to Recent.

Horsetails Spore-bearing plants with segmented stems. Important in the Fossil Record, especially Giant Horsetails which were abundant in the Devonian and the Carboniferous. Horsetails have been declining since then, and only one genus survives today. Range: Devonian to Recent.

ice age A time during which continental ice sheets and glaciers cover large areas of land.

Ichthyosaurs Marine Reptiles with Fish-like bodies and paddle-like limbs, particularly abundant in the Jurassic. Range: The Mesozoic Era.

igneous rocks Rocks formed by the crystallisation of molten magma.

ignimbrite A type of igneous rock formed by the welding together of hot lava materials from an explosive volcanic eruption.

Insects Arthropods with six legs. Range: Devonian to Recent.

intrusion Igneous rock which has forced its way among pre-existing rocks, in the form of batholiths, dykes, sills, etc.

Invertebrates Animals without backbones.

jet stream A stream of very strong wind moving round either of the

Earth's Poles, normally from West to East at altitudes of 10,000 to 50,000 feet.

Jurassic Period The Period between the Triassic and the Cretaceous, in the Mesozoic Era. From 208 to 144 million years ago.

kerogen A carbonaceous material from which hydrocarbons derive.

Labyrinthodont Amphibian Type of early Amphibian whose teeth had "labyrinth" indentations in cross-section. Range: Devonian to Cretaceous. (Australia has the youngest known fossil, from the Early Cretaceous, by which time they were extinct elsewhere.)

lamina Leaf blade.

laminated rocks Rocks which are layered like shale.

laterite Residual deposit of iron oxide produced by weathering of rocks (especially iron-rich ones like basalt) under tropical conditions.

Laurasia The Northern Hemisphere supercontinent which existed at the same time as did Gondwana in the Southern Hemisphere.

lava Volcanic magma in molten or near molten state.

lignin Material of which wood cells are made.

lignite Brown coal with high moisture content, intermediate between peat and bituminous coal.

limestone Sedimentary rock, principally of calcium carbonate.

Linnaean System The system of classification of living things into species, genera, familes, orders, classes and phyla.

lithosphere The outer rigid shell of the solid Earth.

lodestone A naturally occurring form of magnetite which is strongly magnetic.

loess An unconsolidated deposit of rock flour (ground by glaciers), blown into dunes by the wind.

low latitudes Tropics and subtropics.

Lycopods Clubmosses, members of a phylum of spore-producing vascular plants. Important in the Fossil Record since the Late Silurian. Giant Clubmosses in swamp forests of the Northern Hemisphere during the

Carboniferous contributed to coal formation. Range: Late Silurian to Recent.

macrofossil   Fossil visible to the naked eye.

magma   Molten fluids highly charged with vapours and gases, generated in the depths of the Earth. Igneous rocks are formed from the crystallisation of magma.

magnetite   A black iron oxide.

Mammal   Hairy, warm-blooded Vertebrate whose young are nourished on milk.

mantle   Section of the Earth's interior between the crust and the outer core.

Marsupial   Mammal which has a pouch in which the foetal young develop after birth at a very early developmental stage.

meiosis   A process of cell division in which the chromosome number is halved.

mesa   Flat-topped, table-like mountain which falls away steeply on at least three sides. Formed by dissection of a plateau in arid regions.

Mesozoic Era   The Triassic, Jurassic and Cretaceous Periods. From 245 to 66 million years ago.

metamorphic rocks   Rocks altered in texture or crystal content by heat and/or pressure.

metamorphism   Change in form in the course of development, as in a Tadpole to a Frog, a Caterpillar to a Moth, and limestone to marble, etc.

Metazoa   Animals with bodies formed by more than one cell. The Protists, which are single celled, and the Sponges which have a primitive two-layer organisation as opposed to the three-layer basic structure of other groups, are omitted from the Metazoa.

microclimate   The local conditions which apply in a specialised habitat, determined by topography and the interactions of the biota and its surroundings. Often different from the prevailing conditions which apply outside the special area.

microfossil   Fossils invisible to the naked eye.

micro-organism   Organism so small that it can be seen only under magnification.

mid-ocean ridge   One of a continuous system of mainly submarine ridges. New oceanic crust is created here.

Miocene Epoch   A division of the Tertiary Period. From 23 to 5 million years ago.

mitochondria   Small organelles associated with intracellular respiration.

Molluscs   Phylum Mollusca includes Gastropods, Bivalves and Cephalopods.

Monocotyledons   Angiosperms whose embryos contain one seed leaf.

Monotreme   The lowest order of Mammals, of which only the Platypus and the Echidna survive today. They are egg-laying and milk-producing.

monsoon   Season when wind blows continually to the land from the sea, bringing heavy rain.

moraine   Deposit of rock fragments carried in or under glaciers and left behind when the ice melts.

mosses   Bryophytes. Non-vascular land-plants.

mudstone   An unlaminated sedimentary rock consisting of clay minerals and other very fine-grained sediments.

mutation   A sudden variation in some inheritable character in a plant or animal.

Mycorrhiza   Symbiotic relationship between Fungi and plant roots, resulting in enhanced nutrient up-take.

Myriapod   A terrestrial Arthropod with many walking legs.

Nannoplankton   Ultra-microscopic planktonic organisms, both zooplankton and phytoplankton, particularly coccoliths, which can be detected under the highest power microscopes. Important in deep-sea (pelagic) sediments.

Natural Selection   The process by which the individual with the characteristics best suited to changing environmental conditions succeeds at the expense of others that lack the special adaptations. (Survival of the fittest.)

Nautiloid   A Mollusc with a straight, curved or coiled shell which is septate. The septae are straight.

Neogene   The Miocene and Pliocene Epochs of the Tertiary Period (in the Cainozoic Era).

nucleus   The organelle in a cell in which the chromosomes and their genetic material are enclosed.

oil shale   A fine-grained dark brown or black shale containing kerogen.

Oligocene Epoch   A division of the Tertiary Period from 36.6 to 23.7 million years ago.

ooze   An ocean floor deposit composed of the shells of microplankton.

opal   A form of amorphous silica.

order   A unit of classification. Phyla are divided into classes and classes into orders.

Ordovician Period   Period of the Palaeozoic Era between the Cambrian and the Silurian. From 505 to 438 million years ago.

organelle   A small body enclosed in a membrane within the cytoplasm of a cell.

orocline   An S-shaped bend in a plate margin.

orogeny   An episode of tectonic activity with mountain-building involving folding, faulting and thrusting, usually related to a destructive plate margin.

Ostracod   A small aquatic Arthropod, having a bivalved calcareous shell (carapace) enclosing the body. Range: from Cambrian to Recent.

Palaeocene Epoch   First Epoch of the Tertiary Period. From 66.4 to 57.8 million years ago.

Palaeogene   The Palaeocene, Eocene and Oligocene Epochs of the Tertiary Period (Cainozoic Era).

palaeogeography   The arrangement of land and sea in the geological past.

palaeomagnetism   The record of the ancient magnetic field with which a rock is coded at the time of its formation.

Palynology   The study of fossil pollen and spores.

Pangaea   The "one Earth", when all landmasses were aggregated into one supercontinent.

Panthalassa   The "one ocean", at the time of Pangaea.

parasite   An organism which derives its sustenance from another living organism.

Period   A division of geological time.

245

Permian Marine Collapse An extinction event at the end of the Permian.

Permian Period The Period between the Carboniferous and the Triassic. From 286 to 245 million years ago. The last Period in the Palaeozoic Era.

Phanerozoic Eon The Eon of visible life. From the Cambrian to the Present.

photosynthesis The process by which plants synthesise organic compounds. The presence of chlorophyll enables them to use solar energy to combine carbon dioxide with the hydrogen of the water molecule (H_2O). Oxygen is released in the process.

phylum (phyla) The major division in the classification of the Plant or Animal Kingdom. Phyla are further divided into classes, classes into orders, etc.

phytoplankton Plankton comprising Algae (plants).

pinnule Division of a leaf.

plankton Aquatic organisms, mainly microscopic, which float or drift in the water.

Plant Kingdom A major division of living things. Plants have the green pigment chlorophyll which enables them to photosynthesise. Thus they form the basis of food chains for 99.9 per cent of all life.

Plate Tectonics The study of the movement of plates of the Earth's crust and the continents which ride on them.

Pleistocene Epoch An Epoch of the Quaternary Period. From 1.6 million to 10,000 years ago, encompassing an ice age.

Plesiosaurs Marine Reptiles of the Mesozoic Era.

Pliocene Epoch An Epoch of the Tertiary Period, from 5.3 to 1.6 million years ago.

plutonic rock Igneous rock formed from magma which has crystallised deep within the Earth's crust.

polarity Direction of the magnetic field.

polycarp Many-seeded fruiting body, as in Cycadophytes.

problematica Organisms of unknown affinity.

pro-Gymnosperms Primitive Seed Plants, ancestral to Gymnosperms.

Prokaryotes Simple cells without nuclei or organelles.

Proterozoic A name given to time before the Cambrian Period.

Protista Kingdom of unicellular organisms.

protoplasm The content of living cells.

Protozoa Single-celled organisms classified in Kingdom Protista (previously classified in the Animal Kingdom).

Pterosaurs Flying Reptiles of the Mesozoic Era.

quartzite A rock consisting almost completely of silica with purely siliceous cement.

Quaternary Period The most recent geological Period — the last 1.6 million years, comprising the Pleistocene and Holocene Epochs.

Radiolarians An order of small Protistans, one of the main constituents of marine plankton. Characterised by internal siliceous skeletons and of surprising beauty. Range: Cambrian to Recent.

Receptaculitids Organisms with very complex structure, assigned to DASYCLYDACEAE. Range: Ordovician to Devonian.

red-beds Rock strata in which iron has been oxidised under tropical conditions to give a red colour.

Reptiles Cold-blooded Vertebrates of the Class Reptilia whose young develop inside an egg which has a tough skin.

reservoir rocks Highly porous and permeable rocks which are able to hold and transmit fluids.

Rhyniophytes Primitive, ancestral vascular plants from which Ferns, Horsetails and all Seed Plants derive.

rifting Process of faulting and tearing between segments as landmasses disintegrate.

rock flour Finely ground sediment produced by glaciers grinding their way across landscapes.

Running Birds Large Birds which cannot fly, including Ostriches, Rheas, Emus and their prehistoric ancestors.

sandstone Sedimentary rock composed mainly of grains of silica.

saprophyte An organism which lives on dead and decaying organic matter.

seafloor spreading The process by which new seafloor is generated on mid-ocean ridges, resulting in the widening of oceans and thus in continents moving further apart.

sedimentary rocks Rocks composed of sediment derived from the erosion of other rocks.

seed Formed from a fertilised ovule. Contains the embryo.

Seed-ferns Extinct seed-bearing plants.

sexual reproduction Reproduction involving the fusion of male and female gametes, each of which has been produced by meiosis and has half the chromosome number of the adults. On fusion the zygote has the full number of chromosomes restored and is the first cell of the embryo.

shale A laminated, fine-grained sedimentary rock (composed mainly of clay minerals) which splits easily on bedding planes.

sill A sheet-like intrusion of igneous rock.

silt Fine sediment.

Silurian Period The Period (in the Palaeozoic Era) between the Ordovician and the Devonian. From 438 to 408 million years ago.

source rocks Sedimentary rocks containing the organic matter from which hydrocarbons are derived.

southern Beech All the *Nothofagus* species (which are confined to the Southern Hemisphere).

Sponges Sessile aquatic animals of the phylum Porifera, having a sac-like body containing spicules. Sponges have a primitive, unspecialised structure which makes them different from the Metazoa.

spore A reproductive cell which develops into the gametophyte in plants that have an alternation of generations in the life cycle.

stomata Pores in the surface of leaves (or some stems) which enable gaseous exchange to take place.

Stromatolites Reef-like structures produced by the activities of Cyanobacteria. Stromatolitic limestones were formed by these organisms. Range: from 3500 million years ago to the present day.

Stromatoporoids Colonial organisms probably related to Sponges, though lacking spicules. They were important reef-building organisms in the Silurian and Devonian Periods. Range: Cambrian to Cretaceous.

symbiosis Organisms living together for mutual benefit.

Terminal Cretaceous Event The extinction event at the Cretaceous to Tertiary time boundary marked by the iridium layer and the disappearance of Dinosaurs.

terranes Pieces of crust with distinctly different geological histories which have come together to form a landmass.

terrestrial deposit A deposit laid down on land as opposed to a marine deposit.

tetrapod A Vertebrate walking on four legs.

Thallophytes Algae. Simple plants without vascular tissues or specialisation.

Thecodonts Early Reptiles which appeared in the Late Permian.

Therapsids One of the groups of Mammal-like Reptiles of the Permian and the Triassic, regarded as ancestral to Mammals.

till Rock formed from glacial sediments.

tombolo A sand spit which connects an island to a coastline.

trace fossils Burrows, tracks, nests, etc, which are preserved in rocks and show that animals were active in the sediments from which they are formed.

Triassic Period Geological Period of the Mesozoic, between the Permian and the Jurassic. From 245 to 208 million years ago.

Trilobites Arthropods. Range: Cambrian to Permian.

Trimerophytes A group of plants forming a link between Rhyniophytes and Higher Plants.

Vascular Plants Those with conducting tissue for the circulation of water and foods through the plant.

Vertebrates Animals of the phylum Chordata with backbones and a cranium protecting the brain. Includes Fish, Amphibians, Reptiles, Birds and Mammals.

xeromorphy Adaptation for drought resistance.

Zosterophylls Primitive land-plants, ancestral to the Clubmosses.

Zygote Cell formed by the fusion of two gametes. The first cell of the embryo. The fertilised egg-cell.

SELECTED BIBLIOGRAPHY AND FURTHER READING

Albani, A.D. & Johnson, B.D. 1974 The Bedrock Topography and Origin of Broken Bay, NSW. *J. Geogr. Soc. Aust.* 21, 2; 209–214.

Albani, A.D. *et al.* 1988 Cainozoic Morphology of the Inner Continental Shelf near Sydney, NSW, *Jnl. Proc. Roy. Soc. NSW* 121, 1–2; 11–28.

Archer, M. & Clayton, G. Eds. 1984 *Vertebrate Zoogeography and Evolution in Australasia.* Hesperion Press.

Augustinus, P. & Colhoun, E.A. 1986 Glacial History of the Upper Pieman and Boco Valleys, Western Tasmania. *Aust. J. Earth Sc.* 33; 181–191.

Barlow, B.A. 1981 The Australian Flora: Its Origin and Evolution. Introduction to *Flora of Australia* 1; 25–75.

Bishop, D.G., Bradshaw, J.D. & Landis, C.A. 1984 Provisional Terrane Map of the South Island, New Zealand. *Proc. Circ. Pac. Terrane Conf.* Stamford.

Brown, D.A., Campbell, K.S.W. & Crook, K.A.W. 1968 *The Geological Evolution of Australia and New Zealand.* Pergamon Press.

Bowler, J.M. 1976 Age, Origins and Expression in Aeolian Landforms and Sediments. *Earth Sci. Rev.* 12; 279–310.

1982 Aridity in the Late Tertiary and Quaternary in Australia, in Barker & Greenslade, Eds., *Evolution of the Flora and Fauna of Arid Australia.* Peacock Publications.

Byrnes, J.G. 1968 Notes on the Nature and Environmental Significance of the RECEPTACULITACEAE. *Lethaia* 1, 4; 368–381.

Clarkson, E.N.K. 1979 *Invertebrate Palaeontology and Evolution.* George Allen & Unwin.

Colhoun, E.A. 1983 Glaciation of the West Coast Range, Tasmania. *Quaternary Res.* 24; 39–59.

1988 Cainozoic Vegetation of Tasmania. *Proc. Int. Palyn. Conf.* Newcastle University.

Colhoun, E.A. & van de Geer, G. 1986 Holocene to Middle Last Glacial Vegetation History at Tullabardine Dam, Western Tasmania. *Proc. Roy. Soc. Lond.* B229; 177–207.

Conway-Morris, S. & Whittington, H.B. 1979 The Animals of the Burgess Shale. *Sci. Amer.* 241, 1; 122–133.

du Toit, A.L. 1937 *Our Wandering Continents:* An hypothesis on Continental Drifting. Oliver and Boyd.

Evans, P.R. 1988 Formation of Petroleum and the Geological History of Australia, in *Petroleum in Australia: The First Century.*

Fleming, C.A. 1962 New Zealand Biogeography. A Palaeontologist's Approach. *Tuatara* 10, 2; 53–108.

1979 *The Geological History of New Zealand and its Life.* Auckland University Press.

Frakes, L.A. 1979 *Climates through Geologic Time.* Elsevier.

Frakes, L.A. *et al.*   1986   Australian Cretaceous shorelines stage by stage. *Palaeogeog. Palaeoclim. Palaeoecol.* 59; 31–48.

Glaessner, M.F.   1984   *The Dawn of Animal Life: a biohistorical study*. Camb. Univ. Pr., Cambridge.

Greenwood, R.M. & Atkinson, A.E.   1977   Evolution of Divaricating Plants in New Zealand in relation to Moa browsing. *Proc. N.Z. Ecol. Soc.* 24; 21–33.

Harland *et al.*   1983   The Geologic Time Scale, in Decade of North American Geology, *Geology*, 504.

Hill, R.S.   1982   The Eocene Megafossil Flora of Nerriga, New South Wales, Australia. *Palaeontographica B* 180, 1–3; 44–77.

   1986   Lauraceous leaves from the Eocene of Nerriga, New South Wales. *Alcheringa* 10; 327–351.

Hill, R.S. & MacPhail, M.K.   1983   Reconstruction of the Oligocene vegetation at Pioneer, North-east Tasmania. *Alcheringa* 7; 281–299.

Hooker, J.D.   1853   The Botany of the Antarctic Voyage of HM DISCOVERY SHIPS *Erebus* and *Terror* in the years 1839–1843. II *Flora Novae Zelandiae*. Pt. 1. Reeve, London.

Keast, A. Ed.   1981   *Ecological Biogeography of Australia*. Junk, The Hague.

Kemp, E.M.   1978   Tertiary Climatic Evolution and Vegetation History in the South-east Indian Ocean Region. *Palaeogeog. Palaeoclim. Palaeoecol.* 24; 169–208.

Kerr, R.A.   1988   No longer willful, Gaia becomes respectable. *Science* 240; 393–395.

Kershaw, A.P.   1981   Quaternary Vegetation and Environments, in Keast, A. Ed. *Ecological Biogeography in Australia*, 81–101. Junk, The Hague.

   1986   A Comparative History of the Vegetation of South-eastern Australia and New Zealand. *Proc. 21st Inst. Aust. Geogr. Conf.* Perth, WA.

   1989   Late Cainozoic Vegetation of Australasia, in Huntley & Webb, Eds. *Vegetation History*. Junk, The Hague.

Lovelock, J.   1979, 1987   *Gaia: A New Look at Life on Earth*. Oxford Univ. Pr.

Martin, H.A.   1978   Evolution of the Australian Flora and Vegetation through the Tertiary. Evidence from Pollen. *Alcheringa* 2; 181–202.

   1981   The Tertiary Flora. In *Ecological Biogeography of Australia*. Keast, A. Ed. Junk, The Hague.

McCulloch, B.   1982   *No Moa*. Canterbury Museum.

Meglitsch, P.A.   1972   *Invertebrate Zoology*. Oxford University Press.

Mildenhall, D.C.   1980   New Zealand Late Cretaceous and Cenozoic Plant Biogeography: A contribution. *Palaeogeog. Palaeoclim. Palaeoecol.* 39; 197–233.

Monastersky, R.   1988   Dinosaurs in the Dark. *Science News* 133, 12; 184–186.

Nanson, G.C. *et al.*	1988	Stratigraphy, Sedimentology and Late Quaternary Chronology of the Channel Country of Western Queensland. In Warner Ed. *Essays in Australian fluvial geomorphology*. Academic Press.
Officer, C.B. *et al.*	1987	Late Cretaceous and Paroxysmal Cretaceous-Tertiary Extinctions. *Nature* 126, 12; 143–148.
Raup, D.M.	1986	*The Nemesis Affair*. A Story of the Death of Dinosaurs and the Ways of Science. Norton.
Raven, P.H. & Axelrod, D.I.	1975	Angiosperm Biogeography and past continental movements. *Ann. Missouri Bot. Gardn.* 61; 539–673.
Retallack, G.J.	1980	Middle Triassic Megafossil Plants and Trace Fossils from Tank Gully, Canterbury, New Zealand. *J. Roy. Soc. N.Z.* 10, 1; 31–63.
	1981	Middle Triassic Megafossil Plants from Long Gully, Otematata, North Otago, New Zealand. *J. Roy. Soc. N.Z.* 11, 3; 167–200.
	1983	Middle Triassic Megafossil marine Algae and land-plants from near Benmore Dam, Southern Canterbury, New Zealand. *J. Roy. Soc. N.Z.* 13, 3; 129–154.
	1985	Triassic Fossil Plant Fragments from shallow marine rocks of Murihiku Supergroup, New Zealand. *J. Roy. Soc. N.Z.* 15, 1; 1–26.
Riek, E.F.	1970	Fossil History. In *The Insects of Australia* (CSIRO). Melb. Univ. Pr.
Rich, P.V. *et al.*	1985	*Kadimakara*. Pioneer Design Studio.
Rich, P.V. & Rich, T.H., *et al.*	1988	Evidence for low temperatures and biologic diversity in Cretaceous High Latitudes of Australia. *Science* 242; 1403–1406.
Rich, P.V. & Rich, T.H. *et al.*	in press	High Latitude Dinosaurs and associated biota from the Early Cretaceous of South-eastern Australia.
Rich, T.H. & Rich, P.V.	1989	Polar Dinosaurs and Biotas of the Early Cretaceous of South-eastern Australia. *Nat. Geogr. Research* 5, 1; 15–53.
Ritchie, A.	1969	Ancient Fish of Australia. *Aust. Nat. Hist.* 16; 218–223.
Ritchie, A. & Gilbert-Tomlinson, J.	1977	First Ordovician Vertebrates from the Southern Hemisphere. *Alcheringa* 1; 351–368.
Romero, E.J. & Archangelsky, S.	1986	Early Cretaceous Angiosperm leaves from Southern South America. *Science* 234; 1580–1582.
Runnegar, B.	1979	Ecology of *Eurydesma* and the *Eurydesma* fauna of the Permian of Eastern Australia. *Alcheringa* 3; 261–285.
Singh, G., Kershaw, A.R. & Clark, R.	1981	Quaternary Vegetation and Fire History in Australia. In Gill, Groves & Noble Eds. *Fire and the Australian Biota*. Aust. Acad. Sci. Canberra.
Skipworth, J.P.	1974	Continental Drift and the New Zealand Biota *N.Z. J. Geogr.* 57; 1–13.

Smith, D.G. Ed.	1981	*The Cambridge Encyclopaedia of Earth Sciences.* Cambr. Univ. Pr.
Sporli, K.B. & Ballance, P.F.	1988	Mesozoic Ocean Floor/Continental Interaction and Terrane Configuration, South-west Pacific region around New Zealand. In Arraham Ed. *The evolution of the Pacific Ocean Margin.* Oxford University Press.
Stevens, G.R.	1974	*Rugged Landscape.* A.H. & A.W. Reed.
	1980	*New Zealand Adrift.* A.H. & A.W. Reed.
Taube, M.	1985	*Evolution of matter and energy on a cosmic and planetary scale.* Springer-Verlag. New York.
Traverse, A.	1988	Plant Evolution Dances to a Different Beat. *Hist. Biol.* 1; 277–301.
Veeevers, J.J. Ed.	1984	*Phanerozoic Earth History of Australia.* Clarendon Press, Oxford.
Webb, L.J.	1978	A structural comparison of New Zealand and South-east Australian Rainforests and their tropical affinities. *Aust. J. Ecol.* 3; 7–21.
Webb, L.J. & Tracey, J.G.	1981	The Rainforests of Northern Australia. In *Australian Vegetation.* Groves, R.H. Ed. Cambr. Univ. Pr.
	1981	Australian Rainforests. Patterns and Change. In *Ecological Biogeography of Australia.* Keast, A. Ed. Junk, The Hague.
Wegener, A.	1929	*The Origin of Continents and Oceans* English translation. Methuen, London.
White, M.E.	1986	*The Greening of Gondwana.* Reed Books, Sydney.

THE GREENING OF GONDWANA

BY MARY E. WHITE

FOSSIL PHOTOGRAPHY BY JIM FRAZIER

The Greening of Gondwana tells the fascinating story of Australia's floral heritage, from its genesis through to the last stages of evolution. It records the changes that have occurred with the arrival of the white man and his crops of exotic plants.

The Fossil Record studied in *The Greening of Gondwana* is not just a collection of dusty old rock specimens with remains of long-dead plants in their substance. It is a tantalising key to visualising the world as it was long ago. The plants tell us of past climates. They conjure up a picture of ancient landscapes, and explain configurations of land and sea on the ever-changing surface of the Earth.

The Greening of Gondwana provides a whole new approach to geoscience for scientists and general readers alike.

331 x 230 mm
256 pages all colour,
Hard cover
ISBN 0 7301 0154 1

INDEX